Nanostructured Materials and Nanotechnology II

Nanostructured Materials and Nanotechnology II

*A Collection of Papers Presented at the
32nd International Conference on Advanced
Ceramics and Composites
January 27–February 1, 2008
Daytona Beach, Florida*

Editors

Sanjay Mathur
Mrityunjay Singh

Volume Editors

Tatsuki Ohji
Andrew Wereszczak

WILEY

A John Wiley & Sons, Inc., Publication

Copyright © 2009 by The American Ceramic Society. All rights reserved.

Published by John Wiley & Sons, Inc., Hoboken, New Jersey.
Published simultaneously in Canada.

For general information on our other products and services or for technical support, please contact our
Customer Care Department within the United States at (800) 762-2974, outside the United States at
(317) 572-3993 or fax (317) 572-4002.

Wiley also publishes its books in a variety of electronic formats. Some content that appears in print may
not be available in electronic format. For information about Wiley products, visit our web site at
www.wiley.com.

Library of Congress Cataloging-in-Publication Data is available.

ISBN 978-0-470-34498-9

Contents

Preface

The 2nd International Symposium on Nanostructured Materials and Nanotechnology was held during the 32nd International Conference on Advanced Ceramics and Composites, in Daytona Beach, FL during January 27-February 1, 2008.

The major motivation behind this effort was to create an international platform within ICCAC focusing on science, engineering and manufacturing aspects in the area of nanostructured materials. The symposium covered a broad perspective including synthesis, processing, modeling and structure-property correlations in nanomaterials. More than 90 contributions (invited talks, oral presentations, and posters), were presented by participants from more than fifteen countries. The speakers represented universities, research institutions, and industry which made this symposium an attractive forum for interdisciplinary presentations and discussions.

This issue contains peer-reviewed (invited and contributed) papers incorporating latest developments related to synthesis, processing and manufacturing technologies of nanoscaled materials including nanoparticle-based composites, electrospinning of nanofibers, functional thin films, ceramic membranes and self-assembled functional nanostructures and devices. These papers discuss several important aspects related to fabrication and engineering issues necessary for understanding and further development of processing and manufacturing of nanostructured materials and systems.

The editors wish to extend their gratitude and appreciation to all the authors for their cooperation and contributions, to all the participants and session chairs for their time and efforts, and to all the reviewers for their valuable comments and suggestions. Financial support from Plasma Electronic GmbH, Neuenburg, Germany as well as the Engineering Ceramic Division of The American Ceramic Society is gratefully acknowledged. Thanks are due to the staff of the meetings and publication departments of The American Ceramic Society for their invaluable assistance.

We hope that this issue will serve as a useful reference for the researchers and

technologists working in the field of interested in science and technology of nanostructured materials and devices.

Sanjay Mathur
University of Cologne
Cologne, Germany

Mrityunjay Singh
Ohio Aerospace Institute
Cleveland, Ohio, USA

Introduction

Organized by the Engineering Ceramics Division (ECD) in conjunction with the Basic Science Division (BSD) of The American Ceramic Society (ACerS), the 32nd International Conference on Advanced Ceramics and Composites (ICACC) was held on January 27 to February 1, 2008, in Daytona Beach, Florida. 2008 was the second year that the meeting venue changed from Cocoa Beach, where ICACC was originated in January 1977 and was fostered to establish a meeting that is today the most preeminent international conference on advanced ceramics and composites

The 32nd ICACC hosted 1,247 attendees from 40 countries and 724 presentations on topics ranging from ceramic nanomaterials to structural reliability of ceramic components, demonstrating the linkage between materials science developments at the atomic level and macro level structural applications. The conference was organized into the following symposia and focused sessions:

Symposium 1	Mechanical Behavior and Structural Design of Monolithic and Composite Ceramics
Symposium 2	Advanced Ceramic Coatings for Structural, Environmental, and Functional Applications
Symposium 3	5th International Symposium on Solid Oxide Fuel Cells (SOFC): Materials, Science, and Technology
Symposium 4	Ceramic Armor
Symposium 5	Next Generation Bioceramics
Symposium 6	2nd International Symposium on Thermoelectric Materials for Power Conversion Applications
Symposium 7	2nd International Symposium on Nanostructured Materials and Nanotechnology: Development and Applications
Symposium 8	Advanced Processing & Manufacturing Technologies for Structural & Multifunctional Materials and Systems (APMT): An International Symposium in Honor of Prof. Yoshinari Miyamoto
Symposium 9	Porous Ceramics: Novel Developments and Applications

Symposium 10	Basic Science of Multifunctional Ceramics
Symposium 11	Science of Ceramic Interfaces: An International Symposium Memorializing Dr. Rowland M. Cannon
Focused Session 1	Geopolymers
Focused Session 2	Materials for Solid State Lighting

Peer reviewed papers were divided into nine issues of the 2008 Ceramic Engineering & Science Proceedings (CESP); Volume 29, Issues 2-10, as outlined below:

- Mechanical Properties and Processing of Ceramic Binary, Ternary and Composite Systems, Vol. 29, Is 2 (includes papers from symposium 1)
- Corrosion, Wear, Fatigue, and Reliability of Ceramics, Vol. 29, Is 3 (includes papers from symposium 1)
- Advanced Ceramic Coatings and Interfaces III, Vol. 29, Is 4 (includes papers from symposium 2)
- Advances in Solid Oxide Fuel Cells IV, Vol. 29, Is 5 (includes papers from symposium 3)
- Advances in Ceramic Armor IV, Vol. 29, Is 6 (includes papers from symposium 4)
- Advances in Bioceramics and Porous Ceramics, Vol. 29, Is 7 (includes papers from symposia 5 and 9)
- Nanostructured Materials and Nanotechnology II, Vol. 29, Is 8 (includes papers from symposium 7)
- Advanced Processing and Manufacturing Technologies for Structural and Multifunctional Materials II, Vol. 29, Is 9 (includes papers from symposium 8)
- Developments in Strategic Materials, Vol. 29, Is 10 (includes papers from symposia 6, 10, and 11, and focused sessions 1 and 2)

The organization of the Daytona Beach meeting and the publication of these proceedings were possible thanks to the professional staff of ACerS and the tireless dedication of many ECD and BSD members. We would especially like to express our sincere thanks to the symposia organizers, session chairs, presenters and conference attendees, for their efforts and enthusiastic participation in the vibrant and cutting-edge conference.

ACerS and the ECD invite you to attend the 33rd International Conference on Advanced Ceramics and Composites (http://www.ceramics.org/daytona2009) January 18–23, 2009 in Daytona Beach, Florida.

TATSUKI OHJI and ANDREW A. WERESZCZAK, Volume Editors
July 2008

ONE-DIMENSIONAL NANOSTRUCTURED CERAMICS FOR HEALTHCARE, ENERGY AND SENSOR APPLICATIONS

S Ramakrishna[1, 2, 3*], Ramakrishnan Ramaseshan[1, 2], Rajan Jose[1], Liao Susan[3], Barhate Rajendrakumar Suresh[1], Raj Bordia[4]

[1]*NUS Nanoscience and Nanotechnology Initiative, 2 Engineering Dr 3, Singapore 117576*
[2]*NUS Dept of Mechanical Engineering, 9 Engineering Dr 1, Singapore 117576*
[3]*NUS Divn of Bioengineering, 9 Engineering Dr 1, Singapore 117576*
[4]*Dept of Materials Science and Engineering, University of Washington, Seattle, WA 98195*

ABSTRACT:

One dimensional nanostructured materials possess a very high aspect ratio and consequently they possess a high degree of anisotropy. Coupled with an extremely high surface area, this leads to an interesting display of properties in the one-dimensional nanostructured ceramics, which differ markedly from their bulk counterparts. These characteristics have made the one-dimensional nanomaterials to be most sought in mesoscopic physics and in fabrication of nanoscale, miniaturized devices. Electrospinning is an established method for fabrication of polymer nanofibers on a large scale. By electrospinning of a polymeric solution containing the ceramic precursor and subsequent drying, calcination, and sintering, it has been possible to produce ceramic nanostructures and this technique appears highly promising for scale-up. During the last five years, there has been remarkable progress in the fabrication of ceramic nanorods and nanofibers by electrospinning. Ceramic nanofibers are becoming useful and niche materials for several applications owing to their surface- and size-dependant properties. In this paper three main case studies will be presented which elucidate the versatility of ceramic nanofibers in the domains of healthcare, renewable energy and sensor applications.

INTRODUCTION

Advanced ceramic materials constitute a mature technology with a very broad base of current and potential applications and a growing list of material compositions. Advanced ceramics are inorganic, nonmetallic materials with combinations of fine-scale microstructures, purity, complex compositions and crystal structures, and accurately controlled additives. Such materials require a level of processing science and engineering far beyond that used in making conventional ceramics.

Advanced ceramics are wear-resistant, corrosion- resistant and lightweight materials, and are superior to many other material systems with regard to stability in high-temperature environments. Because of this combination of properties, advanced ceramics have an

---

[*] Email address for correspondence: sceram@nus.edu.sg

1

especially high potential to resolve a wide number of today's material challenges in process industries, power generation, aerospace, transportation, military and healthcare applications[1].

Nanostructures of advanced ceramic materials are noted for their stability compared to their non-oxide counterparts and find diverse technical applications. The nanostructured ceramic materials could virtually replace all the bulk ceramics due to their high value-addition in applications such as catalysis, fuel cells, solar cells, membranes, hydrogen storage batteries, structural applications requiring high mechanical strength, in biology for tissue engineering, biomolecular machines, biosensors, etc. Besides, nanostructured ceramic oxides have potential applications in advanced optical, magnetic and electrical devices due to the physical properties these materials posses on account of their electronic structure.

One-dimensional nanostructures can be fabricated on a laboratory scale by advanced nanolithographic techniques such as focused-ion-beam writing, X-ray lithography, etc[2]; however, some of these techniques are suited to only a few material systems[3] and moreover development of these techniques for large scale production at reasonably low costs requires great ingenuity[4]. In contrast unconventional methods based on chemical synthesis such as electrospinning might provide an alternative for generation of one-dimensional nanostructured ceramics in terms of material diversity, cost, throughput and potential for high volume production. The field of ceramic nanofibers made via electrospinning is rapidly growing, as seen in Figure 1.[5]

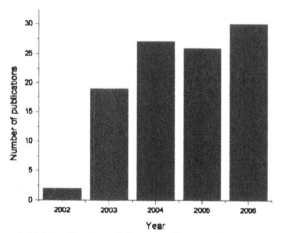

Figure-1: Publication trends in ceramic nanofibers upto 2006[2]

This increasing interest in electrospinning stems primarily from the fact that it is a simple, versatile and relatively inexpensive technique for synthesizing nanofibers. It is precisely the versatility of the technique that has allowed the synthesis of about 40 different ceramic systems[2]. In addition, unlike other methods which produce relatively short nanorods or carbon nanotubes, electrospinning produces continuous nanofibers. This continuity offers the potential for alignment, direct writing, and spooling of the fibers. This potential has been recently demonstrated in several laboratory scales, proof of concept type of experiments.[6]

With the expansion of electrospinning from polymers to composites and to ceramics, the applications for electrospun fibers are vastly expanded across the domains of healthcare, renewable energy and advanced electronics. This paper shall review the advancements made by electrospun nanostructured ceramics across in of these three domains.

I. NANOSTRUCTURED CERAMICS IN HEALTHCARE - BIOMEDICAL APPLICATIONS

Nature bone is a composite comprising 70% minerals mainly in the form of nano-HA and 30% organic matrix mainly in the form of type I collagen. In addition to a network of interconnected micro-pores, bone also has a nanostructure made up mainly of collagen nanofibers and nano-HA. It is increasingly clear that nano-texture plays a significant role in enhancing cell-scaffold interaction. It is therefore desirable that the next generation of bone graft substitutes would incorporate the known composition and structure of natural bone. The objective of this study is to develop bone graft substitute in the form of a three-dimensional (3D) scaffold that not only has the desirable material composition but also bone like micro and nano-texture.

A three-dimensional (3D) scaffold was fabricated using a novel electrospinning setup based on a dynamic liquid support system[3]. Collagen type I, a major organic component of bone and a biodegradable polymer and polycaprolactone (PCL) were used to prepare the scaffold[7]. PCL and PCL/collagen three-dimensional scaffold was mineralized using the alternate soaking[8] and the co-precipitation methods[9].

By electrospinning on a dynamic liquid support, the nanofibers coalesced into bundles of yarn (nanoyarn). The folding of these strands of rope-like nanoyarn creates a 3D scaffold with interconnected micropores. Freeze-dried PCL and PCL/Collagen 3D scaffold were made out of a network of nanoyarn with pore size ranging from a few micrometers to a few hundred micrometers as shown in Figure 2(A). Under SEM, it can be seen that individual yarns from PCL 3D scaffold were made out of aligned nanofibers while PCL/Collagen 3D scaffold were more random.

Alternately dipping the scaffold in $CaCl_2$ and Na_2HPO_4 creates deposition of HA nanoparticles on the 3D PCL/Collagen scaffold as shown in Figure 1(b) while no HA were found on pure PCL scaffold. Nevertheless, some HA were deposited on PCL scaffold by co-precipitatation in $CaCl_2$ and collagen solution as shown in Figure 3 (a). Human fetal osteoblast cells and mesenchymal stem cells (MCS) were seeded on the scaffolds and observed[10].

Human fetal osteoblast cells were found to adhere well to PCL/Collagen 3D scaffold and mineralized PCL/Collagen 3D scaffold (Figure 2 (c)). However, very few cells were found on PCL 3D scaffold. MSCs seeded on PCL that were mineralized by co-precipitation in $CaCl_2$ and collagen solution were also observed to attach well to the 3D scaffold as shown in Figure 2 (b).

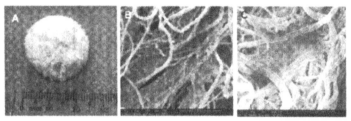

Figure 2. [A] Freeze-dried PCL/Collagen 3D [B] Mineralized PCL/Collagen scaffold with HA (arrows) nanoparticles using alternate soaking method. [C] osteoblast cultured on mineralized PCL/Collagen 3D scaffold after 3 days.

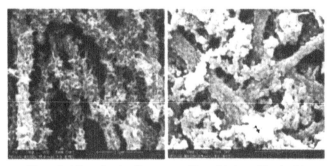

Figure 3: Mineralized collagen nanofibers fabricated by double soaking method, reaction time 5 min (left) and 10 min (right)

Conventional electrospinning is a versatile process for producing sheets of nanofibers from different materials and compositions. By modifying the setup, we were successful in fabricating 3D scaffold made of nanofibrous yarn. The resultant 3D scaffold has interconnected pores of varying sizes. The larger pore of more than 100 μm will be favorable for bone ingrowth and angiogenesis. As pure collagen degrades too rapidly, a blend of PCL and collagen was used to provide structural support during cell migration and proliferation. In our fabrication process, HA that were incorporated onto the nanofibers were in the form of biomineralized HA nanoparticles to resemble the HA found in natural bone. The alternate soaking method has been shown to be a rapid method of depositing HA nanoparticles onto substrates compared to other method of biomineralization. This study also showed that the presence of collagen is vital for the successful deposition of HA nanoparticles. In pure PCL, very limited HA was deposited on the nanofibers while in PCL/Collagen composite, large quantities of HA were deposited. The hydrophobic nature of PCL is not conducive to formation and deposition of HA. Only when a small amount of collagen was added to $CaCl_2$ solution can HA nanoparticles be deposited on pure PCL scaffold. The reactive amino and carboxylic group in collagen provides nucleation site for the formation of HA nanoparticles. In-vitro study using MSC and osteoblasts showed that these cells adhere well to those scaffolds containing collagen. Furthermore confocal microscopy showed the presence of osteoblasts at the interior of these scaffolds. In spite of the presence of nanotexture, 3-D scaffold made of pure PCL has poor cell adhesion due to the hydrophobic nature of the polymer.

The next stage is to fabricate 3D scaffolds from other polymers such as polylactic acid and the incorporation of biological molecules such as bone morphogenic protein (BMP).

Outlook

HA deposited composite polymer fibers show a great potential for fabrication of bone grafts. By combining the nanocomposite fibers with of growth factors and drugs which aid in healing process, it will be possible to fabricate a bone graft which can be used for reinforcement to treat multiple fractures and osteoporosis. The application of ceramic nanofibers in the field of biomedical implants is still at its infancy and as indicated before the potential effects and benefits are yet to be quantitatively estimated. Although metal oxide nanoparticles have been suggested for chemotherapy, drug targeting and delivery vehicle and as biosensors still they have not been commercialized because their cytotoxicity remains unknown[11]. These nanoparticles could permeate into tissues and other organs because of their small size. We predict that by using electrospun nanoceramics, this issue can be solved due to their macro scale dimension along one direction; however, detailed tests are required to prove this.

II. CERAMIC NANOFIBERS FOR CLEAN ENERGY SOURCES – EXCITONIC SOLAR CELLS:

One of the major challenges that future generations will face is to find out solutions for the increasing energy needs. This challenge stems from the limitations in the stock of natural fossil fuels. Therefore, search for alternate energy source that are not only renewable but also clean from environmental and other hazards has been initiated worldwide. Photovoltaics (PV) are a promising technology that directly takes advantage of our planet's ultimate source of power – the sun. When exposed to light, solar cells are capable of producing electricity without any harmful effect to the environment or device, which means they can generate power for many years while requiring only minimal maintenance and operational costs.

Existing Solar Cell Technologies

Existing types of solar cells may be divided into two distinct classes: conventional solar cells, such as silicon and III-V p-n junctions, and excitonic solar cells, ESCs. Most organic-based solar cells, including dye-sensitized solar cells (DSSCs) fall into the category of ESCs. In these cells, excitons are generated upon light absorption. The distinguishing characteristic of ESCs is that charge carriers are generated and simultaneously separated across a heterointerface. In contrast, photo-generation of free electron-hole pairs occurs throughout the bulk semiconductor in conventional p-n junction cells, the carrier separation upon their arrival at the junction is a subsequent process. This apparently minor mechanistic distinction results in fundamental differences in photovoltaic behavior. For example, the open circuit photovoltage Voc in conventional cells is limited to less than the magnitude of the band bending (Obi); however, Voc in ESCs is commonly greater than Obi[12]. Solid state p-n junction solar cells made from crystalline inorganic semiconductors (e.g. Si and GaAs) have dominated the commercial PV market for decades. Commercial solar cell modules (area typical 1 x 2 m^2) of efficiency ~ 17% and single cells (area ~1 cm^2) of efficiency up to 40% (under high optical concentration) have been realized from crystalline silicon and multijunction devices, respectively[13]. A thorough documentation of the progress in photovoltaics till the year 2006 can be found in an earlier report[14]. Currently the wide-spread use of photovoltaics over other energy sources is limited by its relatively high cost per kilowatt-hour, however, ESCs are likely to be an exception due to the possibilities of cost-effectiveness and ease of fabrication compared to the crystalline silicon and III-V p-n junction solar cells[8, 10, 15]

Principle of working of a DSSC

The photovoltaic effect in DSSC occurs at the interface between a dye-conjugated photoelectrode and an electrolyte. A DSSC consists of three functional parts (Figure 4a); viz. a solar light harvester, usually a dye, which converts an absorbed photon into an exciton; an electron acceptor (electrode) that splits the exciton into electrons and holes by the energy difference between the LUMO of the light harvester and the conduction band of the electrode; and a redox mixture that injects the electron back to the dye. The final photoelectric conversion efficiency of DSSC depends on many factors.

There are at least nine fundamental processes that can control the final energy conversion efficiency in an excitonic solar cell (Figure 4b). The fundamental processes occur in ESCs are (i) photon absorption which is determined by the wavelength window where the harvester absorbs, intensity of solar radiation at that window, and absorption cross-section of the dye (η_λ); (2) radiative recombination determined by the carrier life time and the probability for radiative recombination in the excited state (τ_{RCR}); (3) exciton diffusion and its diffusion length (v_{DI}) which controlled by the exciton diffusion coefficient (η_{FD}) and exciton life time; (4) interfacial electron transfer and its rate (η_{CI}); (5) interfacial charge recombination determined by the rate at the interface (η_{ICR}); and (6) the exciton relaxation (η_{EBN}) through which the exciton lose its energy due to relaxation; (7) electron transport through the electrode with drift (Ve), (8) the phonon relaxation (η_{PBN}) through which an electron lose its energy via thermalization, (9) the redox potential of the electrolyte and rate of electron transfer to the dye. All these factors are to be clearly understood to achieve high conversion efficiencies.

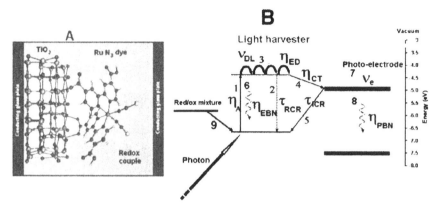

Figure 4: Configuration of DSSC (A). A simplified diagram that shows processes occur in a DSSC. Refer text for definition of the parameters.

Improvement of Conversion Efficiency

Considerable attention was devoted in the past to understand the electrode architecture for efficient electron diffusion and transport[16 17 18 19 20 21 22] as well as choice of electrolytes[23 24 25] and dye molecules[26 27 28 29 30 31] for improving the energy conversion efficiency of DSSC. The best performed DSSC so far produced, which reported an efficiency ~11.1%, utilized a derivative of Ru dye as light-harvester (black dye) and mesoporous TiO_2 as electrode[32]. For outdoor applications, the redox electrolyte containing ionic liquids iodide (I^-) and triiodide (I_3^-) ions are the medium of choice because of their high thermal stability, non-flammability, negligible vapor pressure, and low toxicity. The TiO_2 nanofibers and nanorods recently gained attention for fabrication of DSSC due to the channeled electron transfer in them.[12 13 15] Conversion efficiencies of ~5.8% and ~6.2% are reported in polycrystalline TiO_2 fibers[12] and single crystalline nanorods[15], respectively. Again in both of these cases device working area was rather small (< 0.25 cm^2). Poor adhesion of nanofibers with the conductive glass substrates imposes severe restrictions on the fabrication of large area cost-effective DSSC.

To overcome the adhesion difficulties of TiO_2 nanofibers on conducting glass plates, we developed a technique to fabricate large area electrode layer using electrospun nanofibers[33] Pure anatase TiO_2 nanofibers were prepared by electrospinning a polymeric solution. and subsequent sintering. The details of TiO_2 nanofiber fabrication and their property evaluation are published elsewhere[34]. The electrospun nanofibers were mechanically ground to prepare TiO_2 nanorods. These rods were dispersed in a suitable solvent, spray dried, and sintered to obtain dense electrodes. A schematic of this procedure and final dye-anchored electrodes developed on conducting plate is shown in Figure 5. These dye-anchored TiO_2 nanofibers were used to fabricate large area solar cells. The best-performed DSSC evaluated under AM1.5G (1 sun) condition gave current density ~13.6 mA/cm^2, open circuit voltage ~0.8 V, fill factor ~51% and energy conversion efficiency ~5.8%. We further observed that when dyes are conjugated to nanofibers they showed H-aggregation in contrast to the J-aggregation reported when they are coordinated to TiO_2 nanoparticles. We are currently working on improving the electrical transport properties of nanorod TiO_2 electrodes either by doping with heavy metal ions for quasi-metallic conductivity or by patterning the nanofibers such that the grain boundary scattering are minimized.

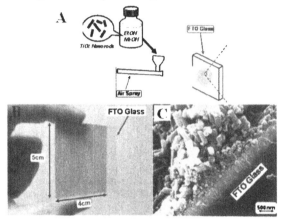

Figure 5: (A) A schematic showing spray deposition of TiO_2 nanorods on the surface of FTO glasses. (B) The TiO_2 nanorods sprayed and sintered on FTO glass. The color of the electrod layer is due to the N3-dye anchoring. The TiO_2 nanorod electrodes were dispersed in a 1:1 vol. mixture of acetonitrile and *tert*-butanol with ruthenium dye ($RuL_2(NCS)_2$-$2H_2O$; L=2.2'-bipyridyl-4,4'-dicarboxylic acid (0.5 mM, N3 Solaronix) for 12 h at room temperature. (C) An SEM image of the spray sintered TiO_2 nanorods.

Performance of solar cells can be improved by introducing highly organized vertically aligned arrays of nanorods as a base of solar cells construction (Figure 6)[15]. High aspect ratio and much bigger, in comparison to classic setup, active area of such structure would increase efficiency and faster ionic and electron mobility along the nanorods would prevent the trapping of electron-hole pairs especially at grains boundaries, what also will find effect in efficiency increase. It is proposed to obtain such structures by electrohydrodynamic shaping of charged solution droplets by longitudinal electric field interaction with them and precisely placing of the created nanorods on the substrate.

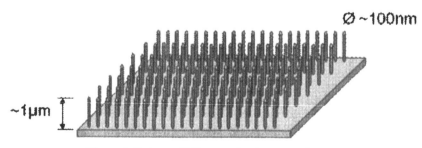

Figure 6: Patterned Electrospinning to produce ordered nanofiber arrays

The Third Generation of Solar Power Harnessing – Application of Quantum Dots

The DSSC has a theoretical limit of conversion efficiency $\sim$ 31%, which could be shifted $\sim$42% if the dyes are replaced by inorganic quantum dots due to the ability of the latter to produce more excitons from a single photon of sufficient energy[36]. This phenomenon is called multi-exciton generation (MEG) or impact ionization and has been demonstrated in quantum dots popular semiconductors such as CdSe and PbSe[37 38 39]. If this property could be exploited to build solar cells, then more fraction of the solar energy could be converted into electrical energy. Colloidal CdS quantum dots were used in DSSC as early as 1990, i.e., within two years from the realization of quantum dots, that gave an energy conversion efficiency $\sim$6%[40]. However this result could not be reproduced and subsequent reports till to date gave efficiencies less than by a factor over two (Table 1). In other words, the performance of quantum dot sensitized solar cells is inferior compared to the conventional dye-sensitized solar cells despite of the remarkable properties of quantum dots.

We recognize that in earlier approaches, quantum dots were simply used as a replacement for dyes without understanding fully the origin and controlling factors of photo-excited electrons and/or their photoelectrochemical properties. For example, the most important requirement for electron injection in DSSC is that LUMO of the light harvester should be at higher energies than conduction band of the photoelectrode. If the LUMO is at similar energies or lower than that of the conduction band of the electrode material, the electron injection is not probable. In conventional photoelectrochemical cells, TiO_2 has been a material of choice as photoelectrode because of its readily availability and relative band positions with many of the dyes. However, the conduction band of TiO_2 is nearly same as that of the conduction band of the bulk CdSe and CdS, two widely studied semiconductors for quantum confinement effect[18]. If these quantum dots are used as light harvesters, the carrier injection is efficient only if the quantum confinement phenomena shift its LUMO to higher levels. This issue is not elaborately addressed in the literature of quantum dots employed photoelectrochemical cells. Further, colloidal quantum dots should be properly attached to organic ligands to bind with the photoelectrodes. These linker molecules have crucial roles in determining the charge transfer and final energy conversion efficiency of photoelectrochemical cells.

We addressed the later problem, i.e., role of a linker molecule on the optoelectronic properties of CdSe quantum dots, recently using experimental results and first principle DFT calculations[41]. This study revealed that oxygen-containing molecules that are conjugated to the surface of CdSe interact strongly with CdSe compared to non-oxygen containing molecules and influence its optoelectronic properties. Our study recommends that surface of quantum dots should be conjugated with proper choice of linker molecules for improving the performance of quantum dot sensitized solar cells.

Outlook

The DSSCs have the potential of producing solar cells of lower cost per kilowatt-hour due to the availability of cheaper ceramic material systems and ease of fabrication compared to the crystalline silicon and III-V p-n junction solar cells. Conversion efficiencies as high as

~11.1% were achieved in DSSC by making use of mesoporous TiO_2 as electrode and black dye. Efforts are currently underway to improve the conversion efficiency by improving the electrical transport properties of nanorod electrodes either by doping with heavy metal ions for quasi-metallic conductivity or by patterning the nanofibers such that the grain boundary scattering are minimized. Besides efforts are also undertook to develop new prototypes of excitonic solar cells in which quantum dots are used as light harvesters in the place of organic dyes. The quantum dots has the potential of increasing the conversion efficiency of solar cells by generating more charge carriers from a single photon of sufficient energy compared to the conventional organic and metallorganic dyes.

III. NANOSTRUCTURED CERAMICS IN SENSOR APPLICATIONS – ELECTROCERAMIC GAS SENSORS

Over the past 20 years, a great deal of research effort has been directed toward the development of gas sensing devices owing to the fact that these sensors have been widely used. Gas sensors are currently used in the following domains-

- The automotive, industrial, and aerospace sector for the detection of NO_x, O_2, NH_3, SO_2, O_3, hydrocarbons, or CO_2 in exhaust gases for environment protection
- The food and beverage industries, where gas sensors are used for control of fermentation processes; and
- The domestic sector, where CO_2, humidity, and combustible gases have to be monitored or detected

The huge variety of applications of sensor technology fuels a continuously growing market, which is expected to exceed \$ 7.5 Billion in 2009 for the USA alone[42]. Some emerging novel applications for these sensors include continuous monitoring of explosive traces which can help to enhance security, monitoring of vapors in medical diagnostics, and in monitoring the level of trace pollutants such as CO, PM10 particles, etc. In a laboratory environment, all these compounds can conventionally be measured using techniques such as IR or UV-Vis spectroscopy, mass spectrometry, or gas chromatography. Although these methods are precise and highly selective, and allow the detection of a single compound in a mixture of gases in very low concentrations, it is obvious that their application is limited by cost, instrumentation complexity, and the large physical size of the instrumentation.

For low-cost and mobile applications, solid-state gas sensors are most common. Such a sensor element has to transform chemical information, originating from a chemical or physical reaction of the gas molecule to be detected with the gas-sensitive material, into an analytically manageable signal. Considerable efforts have been undertaken to develop sensors for these novel applications, however, many of these efforts have not yet reached commercial viability because of problems associated with the sensor technologies applied to gas-sensing systems[43]. Inaccuracies and inherent characteristics of the sensors themselves have made it difficult to produce fast, reliable, and low-maintenance sensing systems comparable to other micro-sensor technologies that have grown into widespread use commercially[44]. With the increasing demand for better gas sensors of higher sensitivity and greater selectivity, intense efforts are being made to find more suitable materials with the required surface and bulk properties for use in gas sensors.

Working principle of a gas sensor

The principle behind solid-state gas sensors is the reversible interaction of the gas with the surface of a solid-state material resulting in a change in material's conductivity. In addition to the conductivity change of gas-sensing material, the detection of this reaction can be performed by measuring the change of capacitance, work function, mass, optical characteristics or reaction energy released by the gas/solid interaction[43]. This principle is illustrated in figure 7 below.

Various materials, synthesized in the form of porous ceramics, and deposited in the form of thick or thin films, are used as active layers in such gas-sensing devices. The read-out of the measured value is performed via electrodes, diode arrangements, transistors, surface wave components, thickness-mode transducers or optical arrangements. However, in spite of so big variety of approaches to solid-state gas sensor design the basic operation principles of all gas sensors above mentioned are similar for all the devices. As a rule, chemical processes, which detect the gas by means of selective chemical reaction with a reagent, mainly utilize solid-state chemical detection principles as shown in figure 7.

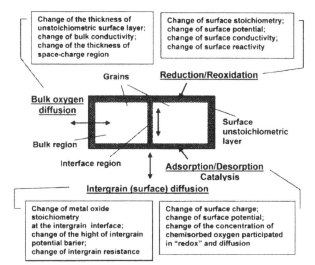

Figure 7: Illustration of the processes that take place in metal oxides during gas detection [43]

Materials used in a sensor

Solid state sensors have been fabricated from a wide variety of materials such as solid electrolytes, classical semiconductors, insulators, metals and organic polymers[45]. The details have been furnished in table 1.

Table 1: Solid State Sensor Materials and Applications

Type of Sensor	Materials	Analyte
Semiconductor based sensors	Si, GaAs	H , O_2, CO_2, H_2S, propane etc.
Semiconducting metal oxide sensors	SnO_2, ZnO, TiO_2, CoO, NiO, WO_3	H_2, CO, O2, H_2S, AsH_3, NO_2, N_2H_4, NH_3, CH_4, alcohol
Solid electrolyte sensors	Y_2O_3 stabilized ZrO_2	O_2 in exhaust gases of automobiles, boilers etc.
	LaF_3, Nafion, Zr $(HPO_4)_2.nH_2O$, $SrCe_{0.95}Yb_{0.05}O_3$	F_2, O_2, CO_2, SO_2, NO, NO_2, and H_2O
Organic semiconductors	Polyphenyl acetylene, phthalocyanine, polypyrrole, polyamide, polyimide	CO, CO_2, CH_4, H_2O, NO_x, NO_2, NH_3, chlorinated hydrocarbons

While many different materials and approaches to gas detection are available, metal oxide sensors remain a widely used choice for a range of gas species. These devices offer low cost and relative simplicity, advantages that should work in their favor as new applications emerge. Metal oxide based sensors are much stable and perform well compared to their polymer counterparts. Moreover it is relatively simple to engineer these ceramic materials to optimize sensor performance.

It has been reported[16] that metal oxide sensors comprise a significant part of the gas sensor component market, which generated revenues of approximately $1.5 Billion worldwide in 1998. Significant growth is projected, and the market should exceed $2.5 Billion by 2010.

Advantages of one-dimensional structures

The different 1-D nanostructure arrangements that have been reported in literature to have the potential in sensing application are summarized in figure 8[43].

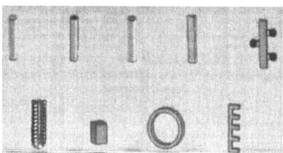

Figure 8: Different 1D metal oxide nanostructures, from top right: nanowire, core-shell structure, nanotubule, nanobelt, dendrite, hierarchical nanostructure, nanorod, nanoring, nanocomb[45]

Oxygen ions adsorb onto the surface of the 1-D nanostructured material, removing electrons from the bulk and creating a potential barrier that limits electron movement and conductivity. When reactive gases combine with this oxygen, the height of the barrier (Schottky) is reduced, increasing conductivity. This change in conductivity is directly related to the amount of a specific gas present in the environment, resulting in a quantitative determination of gas presence and concentration. These gas-sensor reactions typically occur at elevated temperatures (150-600°C), requiring the sensors to be internally heated for maximum response. The operating temperature must be optimized for both the sensor material and the gas being detected. In addition, to maximize the opportunities for surface reactions, a high ratio of surface area to volume is needed. As an inverse relationship exists between surface area and particle size, nano-scale materials, which exhibit very high surface area, are highly desirable. One dimensional nanomaterials have a unique preference for sensor fabrication[47]. This is because of their size dependant behavior. This quantum size effect is reported to be seen in 1-D nanomaterials of size < 50 nm and functions to enhance the sensor properties.

Structural Engineering of materials to enhance sensor performance

Structural engineering of metal oxide films is the most effective method used for optimization of solid state gas sensors. The considerable improvement of such operating parameters as gas response, selectivity, stability, and rate of gas response can be achieved due to optimization of both bulk and surface structure of applied metal oxide films.

Besides the particle size, the influence of the microstructure, that is, the substrate thickness and its porosity, are the other factors that affect response time and the sensitivity. Sensing layers are penetrated by oxygen and analyte molecules so that a concentration gradient is formed, which depends on the equilibrium between the diffusion rates of the reactants and their surface reaction. The rate leading to the equilibrium condition determines the response and recovery time. Therefore, a fast diffusion rate of the analyte and oxygen into the sensing body, which depends on its mean pore size and the working temperature, is vital. Furthermore, maximum sensitivity will be achieved if all percolation paths contribute to the overall change of resistance, that is, that they are all accessible to the analyte molecules in the ambient. Thus a lower substrate thickness together with a higher porosity contributes to a higher sensitivity and faster response time. This was verified experimentally most recently by Yamazoe and coworkers[48,49] investigating the gas response on H_2 and H_2S of thin films of monodisperse SnO_2 with particle diameters ranging from 6–16 nm. It was found that the sensor response was greatly enhanced with decreasing film thickness but with increasing grain size up to 16 nm. The latter appears to be unexpected but can be understood in terms of an increased porosity, which cannot be achieved with the smallest particles studied.

Recently we[50] demonstrated the giant piezo-response from nanofibers of PZT (lead-zirconium titanate) material prepared by the electrospinning technique. It was found that the strain in these one-dimensional nanofibers were 5 times in comparison with their bulk counterpart. Such materials are not only useful as sensing substrate but also in actuators which aid in transduction of the signal into electrical or mechanical response. Application of nanostructured ceramics in actuators could lead to development of devices with higher overall sensitivity and much lower limit of detection could be obtained.

Pure metal oxides are not able to comply with all the demands of a perfect sensing interface. To overcome the inherent limitations of the pure base material, doping with metals and/or oxides has a profound impact on the sensor performance[51]. Note that this process of doping is not comparable with the bulk doping of semiconductors for microelectronic applications. In this case, doping is in fact more the addition of catalytically active sites to the surface of the base material. Ideally, the doping process improves sensor performance by increasing the sensitivity, favoring the selective interaction with the target analyte and thus increasing the selectivity and decreasing the response and recovery time, respectively, which is then accompanied by a reduction of the working temperature. Furthermore, surface doping may enhance the thermal and long-term stability. The control parameters are composition, size, habit, and redox state of the surface modifiers, as well as their dispersion on and/or into the metal oxide surface. As is known from the size-dependent properties of catalytically active nanoparticles, the particle size can effectively control the temperature range as well as the efficiency of a catalytic reaction[52]. The effects of doping are summarized in figure 9.

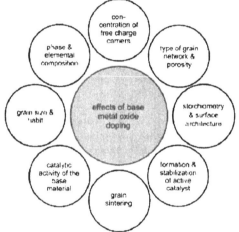

Figure 9: Various effects of metal based doping on ceramic sensor substrates[13]

The modification of LaFeO$_3$ by Sr and Mg had an effect to strongly increase the conductance of the nanocrystalline samples. At low operation temperatures, an exceptional dual conductance response was found at exposure to the tested reducing gases CO, C$_2$H$_4$ and CH$_4$. In the case of CO exposure, a high conductance increase of the LaFeO$_3$ sample was found at low operation temperatures, e.g., by a factor of ~ 300 at exposure to 200 ppm of CO in synthetic air at 100°C. The sensitivity of the LMFO and LSFO samples was much lower than that of the LFO sample at exposure to CO, C$_2$H$_4$ and CH$_4$.[53]

Tan and Zhu[51 55] reported that mixed iron and tin oxide sensors made from nanomaterials that were about 30-50 nm in size showed enhanced sensitivity towards ethanol. The quantum size effect was seen to enhance not only the sensitivity alone but also the selectivity against the target analyte. This property was attributed to the dangling oxygen bonds at the surface which captured the analyte. In fact, the authors reported the selectivity (ethanol Vs

CO and H_2) was 32 times higher than any other ethanol sensing system. Such drastic improvements are possible only through the nanosize.

Outlook

Nano-composites' design is the most promising direction in the development of materials for solid-state gas sensors. One dimensional nanomaterials have recently attracted increasing interest and the possibility of functionalization of these materials with dopants would impart them with unique physical–chemical properties. Highly sophisticated surface-related properties, such as optical, electronic, catalytic, mechanical, and chemical ones can be obtained by advanced nanocomposite 1-D materials, making them attractive for gas sensor applications.

IV. NANOSTRUCTURED CERAMICS IN CATALYSIS AND OTHER APPLICATIONS

Ceramic nanofibers have been explored for other interesting applications such as purification of fuel oils, hot gases and also environmental applications as described in this section. Surface area is the dominant factor that decides the extent of catalysis and its efficiency. Ceramic nanofibers are niche materials which fit this application as they possess extraordinarily high specific surface area. The use of electrospun ceramics for chemical reaction substrates or catalysts is also an upcoming field of interest.

Ceramic membranes can be used to trap heavy metal contaminants from industrial effluents or in rivers. It is well established that the nanosize possesses a high surface activity and this aids in selective capture of contaminants or functional moieties. Taking these advantages into consideration, nanoceramic membranes could be fabricated by electrospinning and used for specialized filtration applications. Moreover by electrospinning it is possible to achieve precise positioning of the fibers and accurate control over pore-size of the catalytic membrane. This leads to fabrication of membranes tailored for a given system to achieve enhanced mass-transfer.

Alumina nanofibers that are about 2 nm in diameter are currently manufactured in the industry[56] (Nanoceram®) for filtration applications. The Nanoceram® alumina fibers are electropositive and attract dust. They also bind and zap viruses and bacteria. The nanosize enables higher flux, atleast two orders of magnitude compared to other membranes and do not get clogged by sub-micron particles which is a major problem in commercial membranes.

Hota et al[57] were the first to prove the concept by fabrication of ceramic nanofibers through electrospinning and showed that it aids in the removal of heavy metal impurities such as Cadmium and Arsenic from waste water. The team found that the removal of heavy metal impurities was enhanced due to the nanostructure configuration of the filter and the capture of contaminants was much better than that of the bulk ceramic material.

Ramaseshan et al[58] have shown the applicability of electrospun ceramic nanofibers as good catalysts for the detoxification of chemical warfare agents. Chemical warfare agents such as the nerve agents (organophosphorus class of compounds) and the blister agents (mustards/sulfides) are quite harmful and are capable of killing or incapacitating the people who are exposed to, even for a short time. These agents are known to react with metal oxides (Mg, Al, Fe, Ti, Zn, Cr, Cu, Mn, etc) which act as catalysts to detoxify them into non-toxic harmless by products. The authors fabricated a mixed metal oxide nanofiber reported as Zinc Titanate and proved their decontamination capacity over warfare simulants such as dimethyl methylphosphonate and chloroethyl ethylsulfide.

Outlook

The catalytic application of bulk phase ceramics has been well recognized explored quite extensively. Miniaturization will lead to improvement of catalytic activity and improve yield and turnover ratios. This potential opens up a wide range of applications for the ceramic nanofibers, from industrial catalysis to filtration, purification and other environmental applications.

SUMMARY

Nanostructured ceramics are indispensable materials for many applications. Although their value-add to several applications has been shown, however only a very few are in commercial form. This is owing to difficulties in their synthesis while maintaining uniform configurations on a large scale. In this review, we have highlighted some important applications wherein the usefulness of nanostructured ceramics have been elucidated and also described how these materials have also been fabricated by electrospinning. Research activities in the field of one-dimensional ceramic nanostructures began only recently and this is still considered as a burgeoning area. Application potentials of ceramic nanostructures in pure form or in hybrid with polymers or other organic materials are still to be explored in detail and require considerable attention by the R&D community.

REFERENCES

[1] Advanced Structural Ceramics, Market Report, published by *BCC Research* (2005)
[2] F Cerrina, C Marrian, *MRS Bull.*, 56, (1996)
[3] Guozhang Cao, Nanostructures and Nanomaterials: Synthesis, Properties and Applications, Imperial College Press, 2003
[4] Y Xia, J A Rogers and G M Whitesides, *Chem. Rev.*, 99, 1823, (1999)
[5] Ramaseshan R, S Sundarrajan, R Jose and S Ramakrishna, *J Appl. Phys.*, **102**, 111101-18, (2007)
[6] Wee-Eong Teo, et al. *Polymer,* **48**, 3400-3405 (2007)
[7] *Solution for electrospinning was prepared by dissolving in 1,1,1,3,3-hexafluoro-2-propanol (HFP, Aldrich Chemical Company, Inc.) and a blend of PCL/collagen type 1 (70:30 w/w) in HFP. Electrospinning was carried out by connecting a high voltage power supply from Gamma High Voltage Research HV. A voltage of 12 kV was applied to the spinneret with the distance from the tip of the spinneret to the surface of the water maintained at 14 cm. The spinneret used was a B-D 27G½" needle which was ground to give a flat tip. A syringe pump was used to provide a constant feed rate of 1 ml/h. The 3D scaffold was formed by allowing water carrying the deposited nanofibers to drain out from the bottom of the basin into a tank, followed by freeze-drying*
[8] Taguchi T et al *Chem. Lett.* **8**, 711-712 (1998)
[9] Liao S et al. *J Biomed Mater Res (Applied Biomater)* **69B**, 158-165 (2004)

[10] *Mineralization procedures for 3D nanofibrous scaffolds*
Alternate soaking method: PCL and PCL/Collagen scaffolds were first immersed in 0.5 M CaCl₂ for 10 min. After rinsed with de-ionized water, the scaffolds were immersed in 0.3 M Na₂HPO₄ for 10 min.
Co-precipitation method: Another sample of PCL was mineralized by co-precipitation in collagen solution in acetic acid, 0.5M CaCl₂ and 0.5M H₃PO₄ solution. 0.5 M NaOH was used to increase the pH value of the solution to 9
Osteoblast and MSCs culture: Human fetal osteoblast cells, hFOB 1.19 (ATCC, US), passage 5, were seeded on 3D scaffolds with a concentration of 3x106 cells/ml. Mesenchymal stem cells (MSCs) were seeded on PCL scaffold mineralized by co-precipitatation in CaCl2 and collagen solution. Culture medium which includes 1:1 mixture of Ham's F12 medium and Dulbecco's modified Eagle's medium without phenol red with 2.5 mM L-glutamine, 0.3 mg/ml G418, and 10% fetal bovine serum was changed every two days. 3D scaffolds with cells cultured for 1 day, 3 days and 7 days were observed by SEM

[11] Nel et al, *Science*, **311**. 5761, 2006, pp. 622 – 627
[12] B. A. Gregg, *J. Phys. Chem.* B 107, 4688-4698 (2003)
[13] M Gräztel, *J. Photochem. Photobio.* **4**, 145-53 (2003)
[14] M. A. Green, K. Emery, D. L. King, Y. Hoshikawa, and W. Warta, *Prog. Photovolt: Res. Appl.* 15, 35-40 (2007)
[15] M. Gratzel, *Nature* 414, 338-344 (2001)
[16] H. Kokubo, B. Ding, T. Naka, H. Tsuchihira, and S. Shiratori, *Nanotechnology* 18 1-6 (2007)
[17] K. Onozuka, B. Ding, Y. Tsuge, T. Naka, M. Yamazaki, S. Sugi, S. Ohno, M. Yoshikawa, and S. Shiratori, *Nanotechnology* 17 1026-31 (2006)
[18] J. B. Baxter and E. S. Aydil, *Solar Energy Materials & Solar Cells* 90. 607-22 (2006)
[19] M. Y. Song, Y. R. Ahn, S. M. Jo, D. Y. Kim, and J. P. Ahn, *Appl. Phy. Lett.* 87 113113 (2005)
[20] M. Law, L. E. Greene, J. C. Johnson, R. Saykally, and P. Yang, *Nature Mater.* 4 455-9 (2005)
[21] M. Y. Song, D. K. Kim, K. J. Ihn, S. M. Jo, and D. Y. Kim, *Nanotechnology* 15 1861-5 (2004)
[22] S. Uchida, R. Chiba, M. Tomiha, N. Masaki, and M. Shirai, *Electrochemistry* 70, 418-20 (2002)
[23] T. Kato, A. Okazaki, and S. Hayase, *J. Photochem. Photobio.* 179 42-8 (2006)
[24] R. Komiya, L. Han, R. Yamanaka, A. Islam, and T. Mitate, *J. Photochem. Photobio.* 164, 123-7 (2004)
[25] P. Wang, S. M. Zakeeruddin, J. E. Moser, M. K. Nazeeruddin, T. Sekiguchi, and M. Grätzel, *Nature Mater.* 2 402-7 (2003)
[26] M. L. Schmidt, J. E. Kroeze, J. R. Durrant, M. K. Nazeeruddin, and M. Grätzel, *Nano Lett.* 5 1315-20 (2005)
[27] M. L. Schmidt, U. Bach, B. R. Humphry, T. Horiuchi, H. Miura, S. Ito, S. Uchida, and M. Grätzel, *Adv. Mater.* 17, 813-5 (2005)

[28] P. Wang. S. M. Zakeeruddin. J. E. Moser, R. Humphry-Baker. P. Comte. V. Aranyos, A. Hagfeldt, M. K. Nazeeruddin, and M. Grätzel, *Adv. Mater.* 16 1806-11 (2004)

[29] T. Horiuchi, H. Miura, and S. Uchida, *J. Photochem. Photobio.* 164 29-32. (2004)

[30] T. Horiuchi, H. Miura. K. Sumioka, and S. Uchida, *J. Am. Ceram. Soc.* 126, 12218-12219 (2004)

[31] M. Adachi, Y. Murata. J. Takao. J. Jiu, M. Sakamoto, and F. Wang, *J. Am. Chem. Soc.* 126. 14943-14949 (2004)

[32] Y. Chiba, A. Islam, Y. Watanabe. R. Komiya, N. Koide, and L. Y. Han. *Jpn. J. Appl. Phys.* Part 2 45. L638 (2006)

[33] K. Fujihara, A. Kumar, R. Jose, S. Ramakrishna, and S. Uchida, *Nanotechnology* 18, 365709 (2007)

[34] A. Kumar, R. Jose, K. Fujihara, J. Wang, and S. Ramakrishna, *Chem Mater.*, 19, 6536-6542 (2007)

[35] The authors would like to acknowledge Dr. Damian Pliszka of NUS Nanobioengineering Labs for his contribution on patterned electrospinning

[36] G. W. Crabtree and N. L. Lewis, *Physics Today* March 2007, 37-40 (2007)

[37] A. J. Nozik, *Physica E: Low-dimensional Systems and Nanostructures* 14, 115-120 (2002)

[38] A. J. Nozik and S. Tetsuo, in Nanostructured Materials for Solar Energy Conversion (Elsevier, Amsterdam, 2006), p. 485-516

[39] M. A. P. R. D. Scaller, V. I. Klimov, , *Appl. Phys. Lett.* 37, 253102 (2005)

[40] R. Vogel, K. Pohl, and H. Weller, *Chem. Phys. Lett.* 174, 241-246 (1990)

[41] R. Jose, M. Ishikawa, V. Thavasi, Y. Baba, and S. Ramakrishna, *J. Nanosci. Nanotechnol.* (In press) (2007)

[42] Zakrzewska, *Thin Solid films*, 391, 229-238, (2001)

[43] Elisabetta Comini, *Analytica Chimica Acta*, 568, 28-40, (2006)

[44] Bong Chull Kim et al, *Sensors and Actuators B*, 89, 180-186, (2003)

[45] Doina Lutic et al, *Topics in Catalysis*, 45 (1), 105-109 (2001)

[46] Y Xia et al, *Adv Mater.*, 15(5), 353-389 (2003)

[47] Harry Tuller, *Sensors and Actuators*, 4, 679-688 (1983)

[48] N Yamazoe et al, *Sensors and Actuators B*, 66 (1-3), 46-48 (2000)

[49] Charles Surya et al, *Appl. Phys Lett.* 91, 113110 (2007)

[50] Zh Zhou, X S Gao, J Wang, K Fujihara, S Ramakrishna and V Nagarajan, Appl. Phys. Lett, 90, 052902, (2007)

[51] P T Moseley, *Sensors and Actuators B*, 6, 149-156 (1992)

[52] Marion E. Franke, Tobias J. Koplin, and Ulrich Simon, *Small*, (1), 36-50 (2006)

[53] Lantto V, et al, *Journal of Electroceramics*, 13, 721-726 (2004)

[54] Zhu et al, *Sensors and Actuators B*, 81, 170-175, (2002)

[55] Tan et al, *Sensors and Actuators B*, 93, 396-401, (2003)

[56] Argonide Corporation Pittsburgh, USA (www.argonide.com)

[57] Hota G, B Rajesh Kumar, Ng WJ and S Ramakrishna, *J Mat Sci*, ce, 43(1), 212-217 (2007)

[58] Ramaseshan R and S Ramakrishna, *J Am Cer Soc*, 90, 1836-42, (2007)

WHAT MAKES A GOOD TiO$_2$ PHOTOCATALYST?

Lars Österlund*, A. Mattsson and P. O. Andersson
FOI, Cementv.20, SE-901 82
Umeå, Sweden

ABSTRACT
    Titanium dioxide photocatalysis is an area which has witnessed an enormous progress during the past three decades. Applications of TiO$_2$ photocatalysis include environmental remediation, self-cleaning coatings, and is also at the heart of TiO$_2$ based energy production (H$_2$ and electricity). Despite an enormous literature a comprehensive understanding of the surface reaction steps on TiO$_2$ is still lacking. This reflects both the complex nature of photocatalytic processes and the difficulties of studying nanoparticles. In this paper we present examples from combined *in situ* molecular spectroscopy studies that highlight the dependence of surface reactions on the structure of TiO$_2$ nanoparticles. We show that for a broad class of organic molecules the reactivity is governed mainly by the bonding and reactivity of a few common intermediate species. The photocatalytic efficiency is correlated with the particle structure and elementary surface reactions steps. We show that μ-formate is a common intermediate that control the overall photo-degradation rate of propane, ketones, and carboxylic acid on rutile TiO$_2$. In contrast, on anatase TiO$_2$ photo-oxidation of acetone is rate determining. This shows that the reactivity of TiO$_2$ is sensitive to both surface modification *and* reactant molecule. Furthermore, the photo-oxidation rate of formic acid depends on the detailed anatase surface properties. This is attributed to a balance of formate bonded to coordinatively unsaturated surface (c.u.s.) Ti atoms and hydrogen bonded molecules due the different bonding strength of formate on c.u.s. sites present on different crystal facets and defects. Ways to improve the surface reactivity of TiO$_2$ nanoparticles are discussed.

1. INTRODUCTION

    The term photodegradation usually refers to the complete oxidation of organic molecules to CO$_2$, H$_2$O and simple inorganic mineral acids that occur under light illumination. For a reaction to be photocatalytic it is further required that the active site (centre) on the catalyst converts a reactant molecule into a product molecule without being consumed itself. This assertion is quantified by the turn over number (TON). The suggested definition of TON is the ratio of the number of photoinduced transformations in a given period of time to the number of photocatalytic sites.[1] A reaction is considered photocatalytic if the turn over number (TON) is greater than unity. Thus whenever TON>1 the active site returns to the original state in the course of the catalytic cycle and initiates a new chemical transformation of a fresh reactant molecule.

    The detailed photocatalytic pathways are in general not known. Yet reported kinetic data may be fitted to simple models borrowed from heterogeneous catalysis such as the Langmuir-Hinshelwood (LH) and Eley-Rideal (ER) type models without definitive knowledge about the particular reaction mechanism.[2,3] Thus agreement with kinetic models is a necessary but not sufficient condition to uniquely attribute kinetic data with a particular reaction mechanism. In fact, many different types of reaction steps may be incorporated in (hypothetical) mechanistic models of photocatalytic reactions and yet give the same kinetic results. However, not discouraged by this fact it is a challenge to disentangle the correct reaction steps, which can pin point rate determining steps and ultimately provide means to modify the photocatalyst or reaction conditions to improve its performance (TON). In the corresponding field of heterogeneous catalysis, basic research employing surface science methods has during the past decades done precisely that by devising simple model systems and methods that capture relevant aspects of real catalysts. This endeavour led Gerhard Ertl to the Nobel

Prize in chemistry in 2007. In fact the surface science of TiO_2 photocatalysis has matured considerably during the past say 10 years. In particularly, our understanding of elementary surface processes on single crystal TiO_2 has advanced considerably, albeit almost exclusively on the rutile surface.[4,5] This is a natural development since gas-solid photocatalysis favourably lends itself to scrutinized studies of well-defined metal oxide systems employing an arsenal of experimental vacuum based techniques, which at the same time provides suitable experimental input to *ab initio* theoretical modelling (without e.g. complications of hydration-shells, solvated ions, etc.). The understanding has thus advanced considerably for simple adsorption systems like O_2, H_2, H_2O, CO, and C_1 and C_2 organic molecules.[4] The implication of these studies are expected to give fundamental insight into heterogeneous photocatalytic processes, and promise to provide a toolbox for predicting reactivity, which is long-sought for in this research field.

Notwithstanding this progress, heterogeneous photocatalysis is complicated by the additional photo-physical properties of the catalyst, photon absorption, transport of photo-excited hot electrons and excitons, and interfacial charge transfer processes. Moreover, there exists a structure gap in photocatalysis, which arises from the fact that most experimental studies have been conducted in colloidal suspensions or on powder samples employing anatase TiO_2, which is commonly considered to be the most active TiO_2 polymorph. On the other hand, much of the fundamental surface science studies conducted during the past decade have been done on rutile single crystals, and in particular the (110) surface, since the most interesting low-index facets of anatase TiO_2 are not readily available. Analogies with the huge body of existing experimental data on TiO_2 nanoparticles are difficult, or even misleading unless a detailed structural analysis of the nanoparticles are done in parallel and translated into the observed reactivity. The literature is not consistent in this respect, and there is a large spread in reported data.

At standard temperature and pressure (STP) conditions TiO_2 has three common polymorphs: rutile (body-centred tetragonal, space group $P4_2/mnm$), anatase (body-centred tetragonal; $I4_1/amd$) and brookite (orthorhombic; *Pbca*).[6] Thermodynamics predicts that rutile is the most stable structure at all temperatures at atmospheric pressures.[6-8] The small difference in the Gibbs free energy between the rutile, anatase and brookite structures indicate, however, that they may coexist at STP conditions. Even though the energy barrier for phase transformation of anatase into rutile structure is low it is kinetically restricted at STP conditions, and is significant only at T>600°C.[9] Taking into account finite crystal sizes variations in surface free energy of exposed facets may reverse the relative thermodynamic stability of the crystal structures when the crystal dimensions become sufficiently small. Indeed, at particle sizes <11 nm anatase is preferred, while at >35 nm rutile is preferred (the larger average surface energy of anatase crystals may explain this reversed stability—see Table 1 below). At intermediate particle sizes brookite is the most stable.[10] Transformation of anatase nanoparticles to rutile must therefore be accompanied by particle size growth (sintering of particles), and a critical particle size has been inferred. Typically the phase transformation of anatase nanoparticles with dimensions exceeding this critical size occurs at T>400°C.[11]

Table 1. Comparison of calculated surface formation energies for relaxed, unreconstructed TiO_2 surfaces. From Refs.[12]

Structure	Surface formation energy (J/m²)
Rutile(110)	0.31
Anatase(101)	0.44
Anatase(001)	0.53
Anatase(100)	0.90
Anatase(110)	1.09

A wide variety of methods have been applied to prepare nanostructured TiO$_2$ employing wet chemical, physical deposition or hybrid methods. They generally yield products with different structures (anatase or rutile), crystallinity, porosity and contaminants. As a consequence the surface properties of the products depend on the preparation technique. Unfortunately, a significant fraction of reported studies involve the TiO$_2$ powder denoted P25 from Degussa AG, which is prepared by vapour-phase oxidation of TiCl$_4$. This process yields a fairly narrow size distribution ($\langle d \rangle$=20-25 nm). However the structure is a mixture of anatase (ca. 75 at.%) and rutile particles and contains chlorine as a main contaminant. Thus the reactivity of P25 cannot easily be related to the surface properties of pure phase TiO$_2$. Chlorine free TiO$_2$ prepared by the sulphate route (usually FeTiO$_3$) contain sulphate ions, which affects its surface acidity. High phase purity and contaminate free TiO$_2$ can be prepared by the alkoxide route (usually Ti(OC$_3$H$_7$)$_4$ or Ti(OC$_4$H$_9$)$_4$). However, both the sulphate and alkoxide routes require a final calcination step to remove water and carbon contaminants, respectively. Since residues and contaminants from the synthesis can affect particle growth and phase transformations (which is particle size dependent—see above), calcination introduces variations in the final product which may be difficult to control. Large progress has been made in various film fabrication methods to prepare TiO$_2$ films with unique properties. In particular physical and chemical vapour phase deposition techniques (PVD and CVD, respectively) allow for high purity, contaminant free crystalline structures without further pretreatment steps (calcination). These latter methods introduce an additional degree of freedom which affects the final TiO$_2$ product, namely the substrate. Interestingly, it has been shown that TiO$_2$ with different growth orientations of surface faces can be obtained compared with conventional wet chemical methods. In particular, it has been reported that TiO$_2$ films prepared by CVD on glass substrates yield structures with preferential (112) oriented crystallites which have beneficial photocatalytic activity.[13] Similarly, surface modified Al/Ti on glass by anodization and subsequent alternating sol-gel dip coating yield preferential (004) oriented crystallites with up to 6 times higher photooxidation rate of acetaldehyde gas.[14] The origin of the beneficial activity of these novel TiO$_2$ films are not known and has been tentatively explained by formation of hollow nanostructures with unusual high exposed surface area (films with both (112)[13] and (004)[14] crystallite orientation), or a high number density of active sites (films with (112) faces[15]). Despite the enormous efforts to synthesize and improve the performance of TiO$_2$ photocatalysts during the past say 30 years, these recent results suggest that we still have plenty of room for improvements and still lack fundamental insight what makes a good TiO$_2$ photocatalyst.

In this paper we take an approach which is a compromise between traditional photocatalysis work and the surface science approach. We take the advantage of the progress of nanofabrication methods to prepare well-defined TiO$_2$ that expose different (distributions of) crystallite orientations. We explore TiO$_2$ prepared by a broad range of physical and wet chemical methods. We review our work employing in situ vibrational spectroscopy to explore at the molecular level the interaction of probe molecules with TiO$_2$. We also include new data which has not been published before. By scrutinizing the adsorption properties and the photochemical reaction pathways of the probe molecules by in situ molecular spectroscopy and relating them to the detailed atomistic structure of the TiO$_2$ obtained by high-resolution transmission electron microscopy (HRTEM), Raman spectroscopy and X-ray diffraction (XRD), we are able to correlate and identify rate determining steps and thus provide independent data (other than kinetic data) that supports a mechanistic interpretation of the observed difference between TiO$_2$ having different structures, crystallinity and contaminants. We exemplify this with the case study of photocatalytic oxidation of propane, where acetone and formate are found to be a key intermediates that limits the oxidation rate.

2. EXPERIMENTAL

2.1 Materials

Anatase and rutile TiO$_2$ nanoparticles were prepared by sol-gel,[16-18] microemulsions,[18,19] and by DC magnetron sputtering[20,21] as described elsewhere. Samples were analyzed either in powder form or as films supported on Si, Al, or CaF$_2$ substrates. The materials were characterized by a variety of experimental methods. Scanning electron microscope images were routinely obtained with a FEG-SEM Leo 1550 Gemini instrument operated at 5kV. Transmission electron microscopy (TEM) was done with a JEOL 2000 FXII instrument operated at 200 kV, and high-resolution TEM with a JEOL JEM 3010 operated at 300 kV (LaB6 cathod). To ensure the sample cleanliness of the titania samples prepared by microemulsions, samples were pressed into thin pellets using a manual hydraulic press (2 tonnes) and analysed by XPS using a Perkin-Elmer PHI 5000C system with Mg K$_\alpha$ radiation (1253.6 eV).[22] The incidence angle was normal to the sample and the detection angle was 45°. In addition to the survey spectra, high-resolution (ΔE=0.025 eV) spectra were collected for the Cl2p, N1s, and Ti2p3 and internally calibrated using the C1s peak as reference. Raman measurements were done with a confocal Raman microscope (Horiba Jobin Yvon LabRAM HR800) operated with a 512 nm Ar ion excitation laser under ambient conditions by placing the samples on aluminium foil and focusing the laser light through a 100× objective (Olympus, NA=0.9 and WD=0.21 mm), or by placing the sample in a water cooled reaction cell (Linkam TS1500) with the possibility to measure at temperatures up to 1500°C and employing a 50× focusing objective (Nikon, SLWD NA=0.45 and WD=17 mm). Reflectance micro-Raman spectra of samples heat treated in air for 30 min at 723 K were recorded with a Renishaw 2000 spectrometer using a 783 nm laser diode light source. Grazing incidence X-ray diffractograms were obtained for samples heat treated at 723 K with a Siemens D-5000 instrument. UV-Vis measurements for films spin-coated on CaF$_2$ substrates were made with a Perkin Elmer Lambda 18 UV-VIS Spectrometer.

2.2 In situ molecular spectroscopy

Photo-induced degradation and surface reactions were measured by in situ Fourier transform infrared (FTIR) spectroscopy by monitoring the time evolution of surface products and adducts. FTIR measurements were made in a vacuum pumped FTIR spectrometer equipped with either an in situ transmission reaction cell, specular cell or a diffuse reflectance cell, which all allow for simultaneous UV illumination, mass spectrometry and in situ FTIR in controlled atmosphere (gas flow, batch mode or vacuum down to 10^{-4} Torr).[17,23] Repeated FTIR spectra were measured with 4 cm^{-1} resolution and typically 135 scans and 30 sec dwell time between consecutive spectra, except for measurements on samples with a silicon substrate, which utilised 90 seconds long dwell time between spectra and a 60 seconds long UV illumination time between spectra. The sample was held at 299 K in a 100 ml min^{-1} gas flow of synthetic air (20% O$_2$ and 80% N$_2$). Prior to each measurement the samples were annealed at 673 K (573 K for the samples on silicon substrates) in synthetic air and subsequently cooled to 299 K in the same feed. Formic acid (GC grade, Merck) and acetone (GC grade, Scharlau) was added to the gas feed through a homebuilt gas generator. The independently calibrated formic acid and acetone concentrations in the gas feed were ca. 7900 ppm and 595 ppm, respectively. No other gases than acetone or formic acid were detected mass spectrometrically. After dosing the sample was kept in the gas feed for ca 15 minutes prior to illumination (except when otherwise stated). FTIR background was collected on a clean sample during 1 minute (265 scans) in synthetic air or N$_2$ feed at 299K.

Thin films or powder materials were irradiated with simulated solar light generated by a Xe arc lamp source operated at 200W employing air mass filters AM1.5 described elsewhere.[17] Briefly light was admitted to the reaction cells through a quartz fibre bundle and focussed through a lens onto the sample surface such that the whole film or powder surface was illuminated (typically 7-10 mm illuminated area). The photon power on the sample was determined to be 166 mW cm^{-2} for wavelengths between 200 and 800 nm, corresponding to ca 12 mW cm^{-2} for $\lambda < 390$ nm.

3. RESULTS AND DISCUSSION

3.1 Deacon's rules and carboxylic acid adsorption

A surface can be considered as a giant bulk defect. At the surface atoms appear which have lower coordination than those in the bulk. These are called coordinatively unsaturated surface (c.u.s) atoms and generally are more reactive than the corresponding saturated atoms, since their valence number make them prone for additional coordination. The c.u.s. Ti^n cations at the TiO$_2$ surface are acidic (Lewis acids sites) and want to form bonds with basic molecules, while the c.u.s. O^{n-} anions are basic (Lewis bases) and adsorb acidic molecules. Classical molecules to probe acidic surface sites on TiO$_2$ are ammonia, pyridine or weak bases such as CO.

It has proven less straightforward to probe the basic surfaces sites of oxides. Davydov et al have suggested that reactive adsorption of CO$_2$ forming surface carbonates is a suitable molecule to probe the basic properties of oxide surfaces.[24] In particular, the splitting of the asymmetric and symmetric carbonate stretching vibrations, $\Delta v_{as-s} \equiv v_{as}(COO) - v_s(COO)$, of the carbonate molecule has been correlated to the strength of the basic sites. For example, the M-O covalency of a lattice O atom involved in monodentate coordinated carbonate is lower than that for a bidentate carbonate. According to Nakamoto,[25] which was first to propose Δv_{as-s} as a measure to distinguish between mono- and bicoordination in metal complexes, Δv_{as-s} is about 300 cm^{-1} for bidentate carbonate. This value *increases* with increasing covalent character of the M-O bond, i.e. decreasing basicity of the oxygen atom. Since the surface atoms on TiO$_2$ nanoparticles are expected to deviate from stoichiometry to a larger degree than bulk terminated samples, e.g. with excess oxygen possessing different valency, the above considerations predict that Δv_{as-s} for CO$_2$ adsorbed such that it forms a bidentate carbonate molecule should be higher if it is coordinated to a three-fold coordinate O atom than a two-fold O atom. Similarly, the surface properties should be influenced by O adatoms with comparably strong basic character.

In analogy with metal-carbonate coordination, Deacon et al have analyzed Δv_{as-s} and the positions of $v_{as}(COO)$ and $v_s(COO)$ for a large number of acetato complexes.[26] By comparing Δv_{as-s} of the free aqueous state ("ionic") and metal coordinated complexes ("metal coordinated"), respectively, they observed the following correlations (Deacon's rules):

Δv_{as-s} (metal coordinated) > Δv_{as-s} (ionic): monodentate coordination

Δv_{as-s} (metal coordinated) < Δv_{as-s} (ionic): bidentate chelating or bridging coordination

Δv_{as-s} (metal coordinated) << Δv_{as-s} (ionic): bidentate chelating coordination

The first correlation has strong experimental support as well as its converse, i.e. monodentate coordinated complexes always exhibit Δv_{as-s} (metal coordinated) > Δv_{as-s} (ionic). However, the inverse of other two correlations are not always obeyed, and it is not straightforward to distinguish the bidentate chelating from the bridging coordination solely based on the magnitude of Δv_{as-s}. In general, these empirical rules should be complemented by additional structural data obtained by independent and theoretical calculations for correct structural assignments. This was also concluded in a recent spectroscopic and theoretical study of formate and acetate adsorption on (large) rutile TiO$_2$ nanoparticles.[27] Moreover, when studying adsorption of carboxylic acid on oxide surfaces interpretation of spectra may be complicated from a practical viewpoint by formation of surface esters and condensation products of carboxylate ions, since these have spectral features close to the carboxylate surface complexes. In this latter case it should also be recognized that the acids can form stable surface complexes which can modify the surface.[24] For example, formic acid adsorption on the rutile (110) surface results in two forms of bridging formate molecules.[18,19,28-30] Upon adsorption the molecule is deprotonated and the H atom transferred to an adjacent (basic) bridging row O atom forming a hydroxyl. The formate ion forms either a complex with two five-fold coordinated c.u.s. surface Ti atoms (μ-formate A), or a complex with an O atom bonded to five-fold coordinated c.u.s. surface Ti atom and one O atom in the position of a lattice O vacancy (Figure 1). The latter complex thus fills the O vacancy site with an oxygen and completes the bridging row structure along the (001) direction.

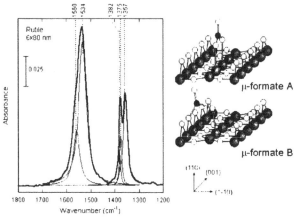

Figure 1. Infrared spectra of formate adsorbed on large rutile particles exposing a large fraction of 110 faces (>95%) along with schematic drawings of suggested surface structures of the adsorbed formate ions. Adapted from Ref. [18,19] (Copyright: The American Chemical Society, 2007).

Despite these difficulties we argue that carboxylic acid adsorption on TiO₂ nanoparticles, and formic acid in particular, is an important and informative tool to probe the surface properties (basic sites) *if* the detailed structure of the nanoparticles are known and complementary data of the adsorbate can be obtained. Moreover, as we will show below the photo-reactive properties for differently coordinated formate molecules are different, which provide further information about their character. Finally, and most important from a practical view point, carboxylic acid is a common intermediate in oxidation of alkanes, alcohols, aldehydes, etc. Since formate (and bridging format in particular) is strongly bonded to TiO₂, oxidation of formate is often a rate determining step for complete oxidation of organic molecules in absence of water to dissolve these species. This is of considerable importance when TiO₂ is used in e.g. air cleaning devices.

3.2 Formic acid adsorption

In the following we give examples on gas phase formic acid adsorption on a variety of different TiO₂ nanoparticles with different structure (anatase, rutile or mixed phase), different particle size and different orientation of their crystal faces. We employ gas phase adsorption to eliminate the contribution from water which complicates the situation. Nevertheless in the *in situ* experiments discussed below the surfaces are always hydroxylated. They are annealed at 400°C in a flow of synthetic air and cooled down in the same feed gas. The surfaces should therefore be regarded as oxidized (with additional adsorbed oxygen on the surface) containing hydroxyls and small amounts of water. This is also evidence in the adsorption experiments where small amounts of OH and H$_2$O are displaced when the acid probe molecules are adsorbed. Conversely, it should be realized that in general during photo-oxidation reactions when H$_2$O and OH is produced (see below) the surface properties changes in the course of the reaction and it is non-trivial to define an ambiguous initial state of the photocatalyst, e.g. as being a non-altered state in the course of the catalytic cycle, which is the true definition of a catalyst.

First we compare adsorption on a TiO$_2$ film prepared by DC magnetron sputtering at room temperature with a similar films post-annealed at elevated temperatures (Figure 2). Although the infrared spectra after HCOOH adsorption appear qualitatively similar the corresponding Raman and XRD spectra show that the film prepared at room temperature contains a large fraction of amorphous domains with only small anatase crystalline domains, while those annealed at T>450°C have well-developed crystallinity with preferential (004) orientation.[21] Thus the former film has a high concentration of c.u.s. atoms exposed to the gas phase. We note that after heat treatment the E$_g$ mode in the Raman spectra shifts from ca. 157 cm^{-1} to 145 cm^{-1} upon annealing to 450°C. indicating that the crystallites in the film prepared at room temperature is ca 2-3 nm while in the annealed films well-developed crystalline domains form containing 15 nm large particles.[31] The latter is agreement with a Scherrer analysis of the XRD data (18 nm).[21] This motivates us to inspect the infrared data more closely. Indeed focussing on the ν_{as}(COO) mode we see that it is shifted to higher energy on the amorphous film by 23 cm^{-1} ($\Delta\nu_{as-s}$ increases from ca. 200 to 230 cm^{-1}).

Inspecting the submonolayer regime (bottom spectra) for the amorphous film we can observe ν_{as}(COO) absorption bands at 1573 and 1615 cm^{-1}. With increasing coverage the 1573 cm^{-1} peak increases and a small shoulder due to H-bonded formate at ca. 1725 cm^{-1} appears. On the annealed film we initially observe a spectral signature very similar to the amorphous film. However, with increasing coverage the 1610 cm^{-1} peak increases and becomes the dominant peak at saturation coverage (at room temperature). We thus ascribe this adsorption behaviour as formate first populating the more basic c.u.s. sites associated with the amorphous structure. Only thereafter the crystalline faces are populated (1610 and 1550 cm^{-1}). In fact, as we will see below it turns out that this is a general observation: formate coordination to basic c.u.s. atoms at defect sites (the amorphous phase can be considered to be represented by a large number of defect sites) is energetically the most favourable adsorption structure on the anatase surfaces.

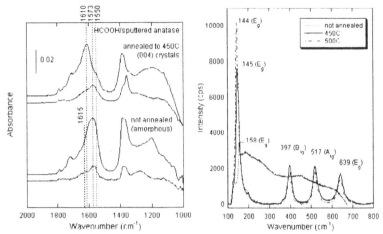

Figure 2. In situ infrared spectra (left) of formate adsorbed on anatase films prepared by DC magnetron sputtering after the indicated different post-annealing temperatures. The film prepared at room temperature is denoted "amorphous". In left panel spectra are shown at submonolayer (bottom spectra) and monolayer coverages (top spectra). In the right panel the corresponding Raman spectra of the pure and annealed films are shown.

In Figure 3 we compare adsorption of formic acid on different well-characterized anatase and rutile nanoparticles.[16-19,23] The nanoparticles expose different crystal faces as determined by HRTEM and XRD (major facets on anatase: (101), (001), (112); major facets on rutile: (110), (101) and (004)). The unique crystal orientations that distinguish each set of nanoparticles are indicated in the figure, although it should be understood that each particle has a distribution of different crystal faces. We include corresponding data for the commercial TiO₂ powders obtained from Degussa (P25) and Merck (BDH). The P25 sample is structurally impure and contains a mixture of anatase (ca 70-80 at.%) and rutile particles, respectively, while the BDH sample contains large pure anatase particles ($\langle d \rangle$=60 nm) exposing mainly (101) faces (Figure 4).

There are several trends to note: (1) Formic acid dissociates on all faces of TiO₂ studied here. (2) Rutile particles with mainly (110) facets (6x80 nm) have a distinct spectral profile characteristic of one particular type of adsorption structure (Figure 1), while all other particles with a more heterogeneous distribution of faces exhibit an equal heterogeneous distribution of formate species, each with its own spectral signature. (3) The adsorption structure on rutile is bridging bidentate (guided by single crystal data.[28-30] and first-principle calculations[17,27,29]). (4) On anatase formate adsorbs first

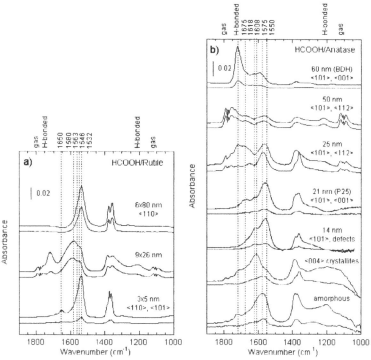

Figure 3. In situ infrared spectra of formate adsorbed on (a) rutile and (b) anatase nanoparticles with different crystallinity and prepared by different methods showing the initial stage of formate adsorption. The bottom spectra for each sample show results at submonolayer coverage. The anatase spectra have been multiplied by a factor of (from top bottom) 0.05, 1.7, 1, 1, 0.1, 0.5, and 0.5, respectively.[18,19]

on c.u.s. atoms associated with defects ($v_{as}(COO) \approx 1570$ cm^{-1}), which we tentatively attribute to a bidentate coordination based on Deacon's rules (whether it is chelating or bridging cannot be deduced based on our data alone). (5) With increasing coverage the basic c.u.s. adsorption sites on the crystal faces becomes populated. Guided by Deacon's rules and the rutile data above we attribute the absorption band at 1550 cm^{-1} to bridging bidentate coordinated to minority anatase faces, since this band is absent on the 60 nm particles with mainly (101) facets. (6) Hydrogen bonded HCOOH are most pronounced on the large anatase and rutile particles exposing large (101) facets. (7) Monodentate formate is associated with absorption bands around 1640-1660 cm^{-1}. (8) The 1610 cm^{-1} peak is correlated to the presence of anatase (004) faces. This peak appears at higher frequency than the corresponding bicoordinated ionic specie, and based solely on Deacon's rule this should therefore be monodentate specie.

Even though these assignments are consistent with the proposed correlation of Ti-O bonding character with the positions of the carboxylate stretching vibrations (note that the Ti-O covalency here does not involve lattice O in contrast to the carbonate case discussed above), we stress that without independent structural knowledge of both the particles and adsorbate the formate adsorption structure cannot be deduced from the infrared data alone. By the comparative approach presented here employing well-characterized nanoparticles we are however able to give qualitative basis for our assertions. For the present purpose it suffices to point out that (i) the relative amount of metal coordinated formate is significantly smaller on the large anatase particles than the small anatase and rutile particles, (ii) μ-formate is the dominant specie on rutile dominated by (110) facets. We conclude that the Ti-O bonding is weaker on the major anatase faces than the rutile faces. Only when minority faces or c.u.s. sites related to defects are present does bicoordination occur on anatase.

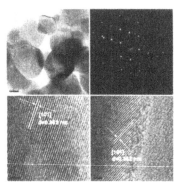

Figure 4. High-resolution TEM and selected area diffraction images of 60 nm anatase particles (BDH). The arrows indicate the (101) lattice spacing in the HRTEM images. The particles have a "Wulff-type" shape (truncated octahedral) that minimizes their surface energy in accordance with Table 1.

3.3 Photodegradation of formic acid

Figure 5 shows the results of photodegradation of formate on TiO₂ films prepared by DC magnetron sputtering. The overall rate of degradation is slightly larger on the crystalline film, but the contribution to this reactivity is very different. On the crystalline film monodentate ($\approx$1655 cm⁻¹) and bridge bonded formate ($\approx$1550 cm⁻¹) decays rapidly, while the main band at 1610 cm⁻¹ is persistent. On the amorphous film the broad peak centred at 1573 cm⁻¹ probably contains a heterogeneous mixture of species. After 48 min illumination species with absorption bands in the region 1573-1600 cm⁻¹ decays slightly faster than the 1610 cm⁻¹ band associated with the (004) facets. The former frequency lies in the region of ionic formate and may indicate that outer sphere formate ions form in the porous amorphous structure. This is qualitatively supported by the comparably large water signal on these samples (not shown). Considering the reported beneficial photocatalytic activity of (004) oriented TiO₂ crystals[14] it may be surprising that formate coordinated to these crystal faces oxidizes more slowly than those present on a highly amorphous TiO₂ film.

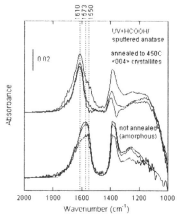

Figure 5. In situ infrared spectra of photodegradation of formate preadsorbed on sputtered TiO₂ films with different structure after different times (0, 6 and 48 min) of solar light illumination employing a 200 W Xe lamp with AM1.5 filters.

In Figure 6 is shown an overview of photodegradation of formate on a wide range of TiO₂ nanoparticles with different structure, crystallinity and size. It is obvious that bridging bidentate on rutile are the most difficult species to photo-oxidize. The degradation rate of formate bonded to c.u.s. atoms associated to defects (e.g. those present on amorphous domains, or on the edges on the small anatase particles, $\leq$25 nm) is also comparably small. In contrast, monodentate formate present in the 1640-1660 cm^{-1} region and bridging bidentate species present around 1550 cm^{-1} (which we have assumed are bonded to minority faces on anatase), swiftly oxidize. Similarly, H-bonded formic acid (like) molecules are rapidly removed from the surfaces of the nanoparticles. The different degradation rate between the 50 nm and 60 nm particles (albeit high on both) is probably due to the outer sphere coordination which reduces the efficiency of surface mediated radical reaction on the latter particles. The water evolution seen on this latter sample qualitatively supports this conclusion. We cannot however exclude that there exist a synergetic effect between the major facets and the minority facets on the 50 nm particles that balances the H-bonding and moderate inner sphere formate coordination. A similar advantageous photo-degradation activity of (112) facets has previously been reported on CVD fabricated anatase films.[13]

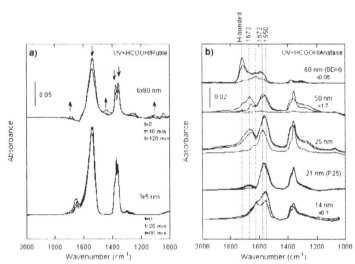

Figure 6. In situ infrared spectra of photodegradation of formate preadsorbed on (a) rutile and (b) anatase nanoparticles after different times (0, 6 and 48 min) of solar light illumination employing a 200 W Xe lamp with AM1.5 filters. The illumination time on the large 6×80 nm particles was extended by 40 min compared to the small 3×5 nm particles to enhance the degradation effect and explicitly show the appearance of the carbonate bands at 1680 and 1100 cm^{-1} (asymmetric) and 1441 cm^{-1} (symmetric). The anatase spectra have been multiplied by the indicated factor.

3.4 Application to the photocatalytic oxidation of propane

The detailed studies of the HCOOH/TiO₂ system presented above are fundamental for a mechanistic understanding of propane photooxidation on TiO₂ nanoparticles. To see why this is so, we recall the reaction pathways for propane oxidation over TiO₂. As reported more than 30 years ago by Teichner et al.[32,33] photocatalytic oxidation of propane proceeds by isopropanol formation, which rapidly converts to form acetone. Acetone is known to be stable and is reacted further only under strongly oxidizing conditions (by C_α-C_β bond cleavage and methyl abstraction[17,34]). The selectivity for aldehydes is in general found to be low. This is not surprising considering that aldehydes are reducing and readily oxidized to carboxylic acid. Protolysis of carboxylic acids to their corresponding ions (e.g. acetic acid to acetate, or formic acid to formate) yield strongly bonded metal coordinated complexes.[25] It is therefore easy to understand why e.g. acetate and formate species are commonly observed on TiO₂. In fact, it has been known since the 1960s that carbonate-carboxylate (R-COO⁻) species form on the surfaces of metal oxides upon reaction with hydrocarbons.[24] Due to the strong bonding of these species to oxides such as TiO₂, it was even thought they irreversibly reacted with the oxide and obstructed the catalytic action. However, it has been established by primarily infrared spectroscopy that R-COO⁻ species are key intermediates in oxidation reactions, which after interaction with O₂ (and O₂ derived species) decompose.[24,35] From this we conclude that we expect that acetone or carboxylate degradation is the rate determining step for propane photooxidation.

We have previously reported[18,23] a similar analysis as the one for formic acid above for acetone/TiO₂. In Figure 7 is shown representative infrared spectra obtained after various times after acetone adsorption on rutile and anatase nanoparticles, respectively. It is obvious that acetone dissociates on rutile particles even in the absence of light illumination. This is in contrast to single crystal data[36] and suggests that surface diffusion to special adsorption sites play a role on the rutile nanoparticles, or that the reduced rutile crystals typically employed in surface science studies behave different than the oxidized TiO₂ employed here . Acetone photo-oxidation proceeds at higher rate than the thermal dissociation pathway, but yields the same surface intermediates as deduced by the same spectral profiles. Taking into account the thermal dissociation on rutile, the photo-oxidation rate on the rutile particles is still 50% higher than on the anatase particles.

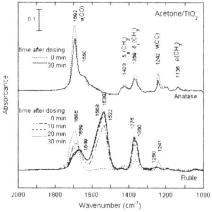

Figure 7. In situ infrared spectra of acetone preadsorbed on 6×80 nm rutile (bottom) and 14 nm anatase (top) nanoparticles at room temperature. Note that acetone dissociate on the rutile particles (in dark). The spectra were acquired in diffuse reflectance mode in an in situ reaction cell with 4 cm⁻¹ resolution.[23]

In the following we combine the knowledge obtained by studying the formate and acetone adsorption systems discussed above to provide a consistent mechanistic reaction pathway for the observed photo-oxidation of propane on anatase and rutile TiO₂ nanoparticles. By monitoring the evolution of surface species on different TiO₂ nanoparticles by in situ infrared spectroscopy and simultaneously measuring the photo-oxidation rate mass spectrometrically, it is possible to work out the momentary carbon mass balance. i.e. the amount of surface species and evolved CO_2 at every different time during the reaction. Since CO_2 is the only gas-phase product in these experiments, the only sources and sinks of carbon are from the gas feed and the surface respectively. In Figure 8 is shown the infrared spectra obtained at different times after illumination on three different types of TiO₂ nanoparticles. The contribution from formate is shown by the grey areas in the figure and represents the Lorenzian curve fits to the measured spectra. It is evident that formate and acetone are the dominant surface species after 45 min illumination on anatase and rutile, respectively. The results in Figure 8 can be converted into a plot of the relative surface coverage (normalized to the initial coverage prior to illumination) versus time where the contributions from the various surface species are resolved. Comparing these data to the measured gas-phase carbon mass balance at each moment during the reaction, the concentration of surface species can be related to the products (CO_2) released into the gas phase. Thus subtracting the evolved CO_2 production from the propane concentration and taking into account stoichiometry (3:1), the concentration of surface species can be deduced from the mass spectrometry data. These data also reveal a lower limit of TON>1 for all particles, and hence show that the reaction is truly photocatalytic on all particles. In Figure 9 is shown a plot of the concentration of surface species deduced from gas-phase mass balance and the concentration measured directly by infrared spectroscopy. The linearity of the curve confirms the correctness of our analysis. Moreover, from Figure 9, we can identify the "best" sample and relate its properties to the physical properties of the particles. The rate determining step on anatase is acetone oxidation. The small anatase particles expose c.u.s. Ti atoms on defects, edges, etc. which forms stronger bonds to the carbonyl and formate groups than on the crystal planes. We tentatively explain the higher oxidation rate (lower momentary surface coverage) on the 21 nm anatase particles compared to the small 14 nm anatase particles with this. On the other hand, on rutile formate degradation leads to strongly bonded bridging formate ions. This is the rate determining step on rutile.[23] Comparing the results for rutile and anatase, we conclude that each particle structure (anatase and rutile) engage in different adsorbate bonding that leads to different rate determining reaction steps (acetone and formate, respectively). We therefore exclude photon-recombination as the primary cause for the variation of the total oxidation rates among the particles, since the intermediate steps preceding or following the RDS do not scale with the total degradation rate. In particular the acetone degradation rate is faster on rutile than on anatase (which is the RDS on anatase but not for rutile). This cannot be explained by a generally larger photon recombination rate on rutile.

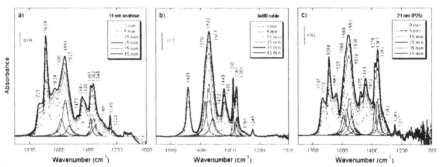

Figure 8. In situ infrared spectra and mass spectra obtained during photocatalytic oxidation of propane over anatase and rutile TiO$_2$ nanoparticles. The spectra were acquired in diffuse reflectance mode in an in situ reaction cell with 4 cm^{-1} resolution.[23] (Reproduced by permission of Elsevier).

There is in additional a large variation among particles with the same modification as evidenced in Figure 6. Preliminary results show that anatase films with (004) facets are superior for propane and acetone photooxidation compared to P25 despite their rather poor ability to degrade formate. This shows that the role of the c.u.s. sites on the (004) facets is to facilitate acetone photo-degradation, which is the rate determining step on anatase. The penalty for this latter structure is that formate degradation is slow (Figure 6), and may explain why anatase particles exposing a distribution of facets (including minority facets) are still better than the (004) films. These results suggest that it is possible to tune the photocatalytic performance by appropriate control (and knowledge) of particle structure for a particular type of reaction. In particular, this work shows that minority facets of anatase probably play a larger role in the observed photo-reactivity of anatase nanoparticles that so far has been discussed in the literature. Moreover, albeit rutile the photocatalytic activity of rutile is commonly considered to be inferior to anatase, our results suggest that this may not necessarily be the case (see e.g. Figure 9). Our results suggest that these differences may be sought for in the Ti-O bond character (bonding strength) in the various the adsorbate-nanoparticle systems rather than in their photochemical properties. However, as we have stressed here, further studies employing e.g. diffraction techniques and first-principle calculations are needed to provide independent complementary data to resolve which types of adsorbate structure that forms on the various minority facets and defective TiO$_2$ surfaces. The in situ vibrational spectroscopy approach reported here on a wide range of TiO$_2$ structures show the potential for combined surface science and synthetic studies to unravel the reactive properties on TiO$_2$ which are instrumental for further developments in this field.

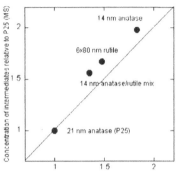

Figure 9. Correlation of gas phase and surface carbon mass balance during photocatalytic degradation of propane over TiO$_2$ nanoparticles.[23] (Reproduced by permission of Elsevier).

4. CONCLUSION

In situ molecular spectroscopy combined with dedicated nanoparticles synthesis provides a powerful approach to study photocatalytic surface reactions on TiO$_2$. Using a reductionistic approach the reaction rate determining steps for propane photo-oxidation has been identified with oxidation of formate (rutile) and acetone (anatase) intermediates. The adsorbate structure on these TiO$_2$ modifications depends on the detailed surface properties and ultimately determines the rate of photodegradation. In particular, the distribution of exposed surface facets is crucial for their reactivity. In the case of formate the infrared frequency of the COO stretching vibrations varies considerably depending on which structure and surface facet of TiO$_2$ formate is adsorbed on. For bicoordination a small value of Deacon's "coordination parameter" Δv_{as-s} indicate a high Ti-O covalency (bonding strength) and a softening of the stretching asymmetric COO mode. On anatase (101) facets formic acid forms hydrogen bonded species, with only a minor fraction bicoordinated to minority facets. Based on data on amorphous TiO$_2$ films and its crystallization, we conclude that on small anatase particles, bidentate coordinated formate forms preferentially at c.u.s. sites associated with defects. On the rutile (101) surface bridging bidentate surface coordination to five-fold coordinated Ti is the most favourable adsorption site. Conversely, photodegradation of formic acid is slowest on particles with a high fraction of (110) facets on rutile of all TiO$_2$ particle structures examined here. In contrast, acetone photo-degradation is faster on the rutile (110) surface than any of the exposed anatase surface facets. The different pathways for propane photodegradation, which is known to proceed by acetone and formate formation, can thus be related to the surface structure of the individual TiO$_2$ nanoparticles. We stress that only by further comparisons with known "reference" facets of TiO$_2$ (surface science studies) and independent structural information, unambiguous assignment of the detailed adsorbate structure be made. Work along these lines is essential for further developments in this field.

ACKNOWLEDGEMENTS
We thank S. Bakardjieva for HRTEM measurements, and M. Leideborg, G. Westin, M. Andersson and A.E.C. Palmqvist for the synthesis of TiO$_2$ nanoparticles.

REFERENCES
[1]A. V. Emeline, V. K. Ryabchuk and N. Serpone, Photoreactions occurring on metal-oxide surfaces are not all photocatalytic. Description of criteria and conditions for processes to be photocatalytic, *Catal. Today*, **122**, 91-100 (2007)
[2]A. V. Emeline, V. K. Ryabchuk and N. Serpone, Dogmas and Reflections in Heterogeneous Photocatalysis. Some Enlightend Reflections, *J. Phys. Chem. B*, **109**, 18515-18521 (2005)
[3]D. F. Ollis, Kinetics of Liquid Phase Photocatalyzed Reactions: An Illuminating Approach, *J. Phys. Chem. B*, **109**, 2439-2444 (2005)
[4]U. Diebold, The surface science of titanium dioxide, *Surf. Sci. Rep.*, **48**, 53-229 (2003)
[5]J. Matthiesen, The influence of point defects on TiO$_2$(111) surface properties, Ph. D. Thesis, University of Aarhus (2007)
[6]A. Earnshaw and N. N. Greenwood, Chemistry of the Elements, Elsevier Science Ltd.: Oxford (1997)
[7]A. Fahmi, C. Minot, B. Silvi and M. Causá, Theoretical analysis of the structures of titanium dioxide crystals, *Phys. Rev. B*, **47**, 11717–11724 (1993)
[8]A. Navrotsky, J. C. Jamieson and O. J. Kleppa, Enthalpy of Transformation of a High-Pressure Polymorph of Titanium Dioxide to the Rutile Modification, *Science*, **158**, 388-389 (1967)
[9]W. W. So, B. P. Park, K. J. Kim, C. H. Shin and S. J. Moon, The crystalline phase stability of titania particles prepared at room temperature by the sol-gel method, *J. Mat. Sci.*, **36**, 4299-4305 (2001)
[10]H. Zhang and J. F. Banfield, Understanding Polymorphic Phase Transformation Behaviou during Growth of Nanocrystalline Aggregates: Insights from TiO$_2$, *J. Phys. Chem. B*, **104**, 3481-3487 (2000)
[11]P. I. Gouma, P. K. Dutta and M. J. Mills, Structural stability of titania thin films, *Nanostruct. Mater.*, **11**, 1231-1237 (1999)
[12]U. Diebold, N. Ruzycki, G. S. Herman and A. Selloni, One step towards bridging the materials gap: surface stuides of TiO$_2$ anatase, *Catal. Today*, **85**, 93-100 (2003)
[13]D. Byun, Y. Jin, B. Kim, J. K. Lee and D. Park, Photocatalytic TiO$_2$ deposition by chemical vapour deposition, *J. Hazard. Mat. B*, **73**, 199-206 (2000)
[14]S.-Z. Chu, S. Inoue, K. Wada, D. Li and H. Haneda, Highly porous TiO$_2$/Al$_2$O$_3$ composite nanostructures on glass by anodization and sol-gel process: fabrication and photocatalytic characterization, *J. Mater. Chem.*, **13**, 866-870 (2003)
[15]S. Tokita, N. Tanaka, S. Ohshio and H. Saitoh, Photo-Induced Reactions of Highly Oriented Anatase Ploycrystalline Films Synthesized Using a CVD Apparatus Operated in Atmospheric Regime, *J. Ceram. Soc. Jpn.*, **111**, 433-435 (2003)
[16]M. Leideborg and G. Westin, Preparation of Ti-Nb-O Nano Powders and Studies of the Structural Development on Heat-Treatment, *Adv. Sci. and Technol.: 9th CIMTEC-World Ceramics Congress and Forum on New Materials*, Florence (1998)
[17]A. Mattsson, M. Leideborg, K. Larsson, G. Westin and L. Österlund, Adsorption and Solar Light Decomposition of Acetone on Anatase TiO$_2$ and Niobium Doped TiO$_2$ Thin Films, *J. Phys. Chem. B*, **110**, 1210-1220 (2006)
[18]L. Österlund and A. Mattsson, Surface characteristics and electronic structure of photocatalytic reactions on TiO$_2$ and doped TiO$_2$ nanoparticles, *Proc. SPIE*, **6340**, 634003 (2006)
[19]M. Andersson, A. Kiselev, L. Österlund and A. E. C. Palmqvist, Microemulsion-mediated room temperature synthesis of high surface area rutile and its photocatalytic performance, *J. Phys. Chem. C*, **111**, 6789 (2007)

[20]C. G. Granqvist, A. Azens, P. Heszler, L. B. Kish and L. Österlund, Nanomaterials for benign indoor environments: Electrochromics for "smart windows", sensors for air quality, and photo-catalysts for air cleaning, *Solar Energy Mater Solar cells*, **91**, 355-365 (2007)

[21]Z. Topalian, J. M. Smulko, G. A. Niklasson and C. G. Granqvist, Resistance noise in TiO₂-based thin film gas sensors under ultraviolet irradiation, *J. Phys.: Conf. Series*, **76**, 012056 (2007)

[22]M. Andersson, L. Österlund, S. Ljungström and A. Palmqvist, Preparation of Nanosize Anatase and Rutile TiO₂ by Hydrothermal Treatment of Microemulsions and Their Activity for Photocatalytic Wet Oxidation of Phenol, *J. Phys. Chem. B*, **106**, 10674-10679 (2001)

[23]T. van der Meulen, A. Mattsson and L. Österlund, A comparative study of the photocatalytic oxidation of propane on anatase, rutile, and mixed-phase anatase–rutile TiO₂ nanoparticles: Role of surface intermediates, *J. Catal.*, **251**, 131-144 (2007)

[24]A. A. Davydov, Molecular Spectroscopy of Oxide Catalyst Surfaces, John Wiley & Sons: Chichester (2003)

[25]K. Nakamoto, Infrared and Raman Spectra of Inorganic and Coordination Compounds, John Wiley & Sons: New York (1997)

[26]G. B. Deacon and R. J. Phillips, Relationships between the carbon-oxygen stretching frequencies of carboxylates and the type of carboxylate coordination, *Coord. Chem. Rev.*, **33**, 227-250 (1980)

[27]F. P. Rotzinger, J. M. Kesselman-Truttman, S. J. Hug, V. Shklover and M. Grätzel, Structure and Vibrational Spectrum of Formate and Acetate Adsorbed from Aqueous Solution onte the TiO₂ Rutile (110) Surface, *J. Phys. Chem. B*, **108**, 5004-5017 (2004)

[28]B. E. Hayden, A. King and M. A. Newton, Fourier Transform Reflection-Absorption IR Spectroscopy Study of Formate Adsorption on TiO₂(110), *J. Phys. Chem. B*, **103**, 203-208 (1999)

[29]Y. Morikawa, I. Takahashi, M. Aizawa, Y. Namai, T. Sasaki and Y. Iwasawa, First-Principles Theoretical Study and Scanning Tunneling Microscopy Observation of Dehydration Process of Formic Acid on a TiO₂(110) surfcae, *J. Phys. Chem. B*, **108**, 14446 -14451 (2004)

[30]S. A. Chambers, M. A. Henderson and Y. J. Kim, Chemisorption geometry, vibrational spectra, and thermal desorption of formaic acid on TiO₂(110), *Surf. Rev. Lett.*, **5**, 381-385 (1998)

[31]S. Kelly, F. H. Pollak and M. Tomkiewicz, Raman Spectroscopy as a Morphological Probe for TiO₂ Aerogels, *J. Phys. Chem. B*, **101**, 2730-2734 (1997)

[32]M. Formenti, F. Juillet, P. Meriaudeau and S. J. Teichner, Heterogeneous Photocatalysis for Partial Oxidation of Paraffins, *Chem. Technol.*, **1**, 680 (1971)

[33]N. Djeghri, M. Formenti, F. Juillet and S. J. Teichner, Photointeraction on the surface of titatnium dioxide between oxygen and alkanes, *Faraday Discuss. Chem. Soc*, **58**, 185 (1974)

[34]J. Cunningham and B. K. Hodnett, Kinetic studies of secondary alcohol photo-oxidation on ZnO and TiO₂ at 348 K studied by gas chromatographic analysis, *J. Chem. Soc., Faraday Trans. 1*, **77**, 2777-2783 (1981)

[35]G. Busca and V. Lorenzelli, Infrared Spectroscopic Identification of Species Arising from Reactive Adsorption of Carbon Oxides on Metal Oxide Surfaces, *Mater. Chem.*, **7**, 89-126 (1982)

[36]M. A. Henderson, Acetone Chemistry on Oxidized and Reduced TiO₂(110), *J. Phys. Chem. B*, **108**, 18932-18941 (2004)

MANUFACTURING OF CERAMIC MEMBRANES CONSISTING OF ZrO_2 WITH TAILORED MICROPOROUS STRUCTURES FOR NANOFILTRATION AND GAS SEPARATION MEMBRANES

Tim Van Gestel, Wilhelm A. Meulenberg, Martin Bram, Hans-Peter Buchkremer
Forschungszentrum Jülich GmbH, Institute of Energy Research, IEF-1: Materials Synthesis and Processing, Leo-Brandt-Strasse, D-52425 Jülich, Germany

ABSTRACT

The preparation and characterization of novel ceramic multilayer microporous membranes with Y_2O_3-doped ZrO_2 functional layers is reported. During preparation, special care is given to each sub-layer of the multilayer configuration: the substrate, the macroporous interlayer, the mesoporous interlayer and the microporous toplayer. Macroporous layers with a pore size of ~ 100 nm are made starting from a commercially available $8Y_2O_3$-ZrO_2 powder. For the deposition of mesoporous and microporous membrane layers, simple scalable sol-gel coating procedures have been developed. In the first part, rather thick standard alumina (γ-Al_2O_3) and novel $8Y_2O_3$-ZrO_2 membrane films are made by dip-coating with sols containing colloidal particles (average particle size 30 - 65 nm). The layers show a smooth surface, a pore size of ~ 5 nm and a thickness of ~ 4 μm and ~ 1 μm, respectively, and can be used as a carrier for an ultra-thin nano-structured membrane layer. In the second part, it is demonstrated that a nano-particle sol can be used for the manufacturing of such a layer made of $8Y_2O_3$-ZrO_2, when the pore size of the mesoporous membrane is adapted to the sol particle size (particle size ~ 6 nm). SEM indicated a thickness of ~ 50 - 100 nm for $8Y_2O_3$-ZrO_2 toplayers obtained by means of a simple dip-coating – calcination procedure. N_2-adsorption/desorption measurements showed a pore size of ~ 1 nm for the membrane material.

INTRODUCTION

Nanofiltration (NF) and gas separation (GS) membranes can generally be classified into two major groups according to their material properties: organic polymeric membranes and inorganic microporous ceramic membranes. Polymeric membranes constitute the most important group and have been commercially available for many years. They are relatively easy to prepare and can be produced cheaply at large scale. However, their application is limited to moderate temperatures and to feed streams which are not too corrosive.

The goal of our current work is to prepare a novel chemically and thermally stable microporous ceramic membrane with an improved pore size and quality than the membranes described in literature. Substantial progress has already been made toward development of NF membranes comprising a functional TiO_2 toplayer, for application in corrosive media. For the best membranes, a pore size of ~ 1 nm has been reported and these membranes can separate effectively small molecules from water or an organic solvent, based on a molecular sieving mechanism [1].

For the separation of mixtures of gases, which show a significantly smaller diameter than the molecules involved in a molecular nanofiltration process, membranes with a smaller pore size in the range of ~ 0.4 - 0.6 nm are required (e.g. kinetic diameter CO_2 0.33 nm, H_2 0,29 nm, N_2 0.36 nm). Membranes with a toplayer made of microporous SiO_2 have been frequently considered for this application, because the material can be rather easily synthesized with the desired pore size, leading to a high selectivity in combination with a relatively high flux. Another important advantage includes the possibility to apply a relatively simple coating method based on common sol-gel technology for the deposition of ultra-thin layers, which is also applicable for the development of membranes at large-scale [2,3].

A significant drawback of current microporous GS membranes, which have been introduced as competitors for polymeric membranes, is, however, the limited chemical stability of the applied membrane materials towards water (vapour), acids and bases. In order to improve the stability towards water, SiO$_2$ membranes with ZrO$_2$ or TiO$_2$ added as a second component have already been developed [4]. In our work, membranes made from Y$_2$O$_3$-doped ZrO$_2$ are proposed. Membrane materials based on zirconia are generally recognized for their excellent chemical stability and – with the addition of a doping compound – also a high thermal stability can be obtained for such materials [5]. Unfavorable layer formation and the difficulty in preparing crack-free toplayers with the required small pore sizes is however always experienced as a major problem during the development of zirconia based membranes. An additional problem includes the strong tendency for cracking in such layers which can not be solved with the addition of binders or other large molecular weight organic additives, because these compounds create large voids in the coated ultra-thin layer after the firing process and prevent the formation of the required very fine microporous structure.

EXPERIMENTAL

1. Preparation of membrane substrate

The first substrate type consists of a porous 8Y$_2$O$_3$-ZrO$_2$ plate with an average pore size of ~ 1 µm and a macroporous 8Y$_2$O$_3$-ZrO$_2$ layer with an average pore size of ~ 100 nm. The substrate plates were prepared according to a large-scale procedure, which is applied in our institute for the manufacturing of solid oxide fuel cells. The first step in the preparation procedure includes the formation of a plate with a size of 25 x 25 cm^2 and a thickness of ~ 1 mm by a warm-pressing procedure. In order to reduce the roughness and the pore size for further modification with very thin mesoporous and microporous membrane layers, an intermediate macroporous 8Y$_2$O$_3$-ZrO$_2$ layer is deposited. In this work, the intermediate layer is made by means of vacuum slip-casting or by screen-printing, starting from a well-known commercially available 8Y$_2$O$_3$-ZrO$_2$ powder (Tosoh corporation, TZ-8Y). Figure 1 shows the particle size measured for a representative suspension of the 8Y$_2$O$_3$-ZrO$_2$ powder by means of a dynamic laser light scattering technique (Horiba LB-550). Consolidation of the macroporous layer was accomplished by sintering in air at 1100°C for 2 h.

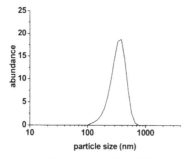

Fig. 1. Particle size distribution of 8Y$_2$O$_3$-ZrO$_2$ suspension used for making a macroporous membrane layer on warm-pressed substrate

The second substrate type was made by vacuum-casting the same suspension into disks with a thickness of ~ 3 mm. Each disk was sintered at 1200°C for 2h and then subjected to surface grinding with a diamond grinding wheel and polished very carefully with diamond particles (6 µm, 3 µm), which left a very smooth nearly mirror-like surface. The final size and thickness of the substrates was ~ 4 cm and ~ 2 mm, respectively.

2. Preparation of mesoporous intermediate membrane layers

Mesoporous membrane layers were made by a colloidal sol-gel coating procedure. In a first coating experiment, alumina membrane layers were prepared from sols containing γ-alumina colloidal particles with a size of ~ 30 nm. The sol preparation was based on the well-known Yoldas process, which includes hydrolysis of a metal-organic precursor ($Al(OC_4H_9)_3$, Sigma-Aldrich) with H_2O and subsequent destruction of larger agglomerates with HNO_3 at elevated temperature (> 80 °C) [6]. In a second coating experiment, a zirconia sol with a size of ~ 30 nm was prepared in a similar way by hydrolysis of $Zr(OC_3H_7)_4$ (Sigma-Aldrich). Yttria-doped zirconia sols (8 mol% yttria) were prepared by adding the proper amount of $Y(NO_3)_3.6H_2O$ to the zirconia sol [7].

Dip-coating experiments were performed using an automatic dip-coating device, equipped with a holder for 4 x 4 cm^2 substrates. The substrates were cut from the porous $8Y_2O_3$-ZrO_2 plate, made by the standard production process. In some cases, orientating coating experiments were first done on the second polished substrate type mentioned above. Coating liquids were prepared from the sols by adding polyvinyl alcohol (PVA) as coating and drying controlling additive. In the coating process, sol particles were deposited as a membrane film by contacting the upper-side of the substrate with the coating liquid (dip-coating), while a small under-pressure was applied at the back-side. In order to obtain the mesoporous γ-Al_2O_3 membrane, firing (calcination) was performed in air at 600°C for 3 h, with a heating and cooling rate of 1°C/h. In the case of the $8Y_2O_3$-ZrO_2 interlayer, the temperature was set at 500°C for 2h. Unless stated otherwise, the entire dip-coating – drying – firing cycle was carried out twice for forming both types of interlayers. For characterization of the material properties (pore size (N_2-adsorption/desorption), phase structure (XRD)), unsupported gel-layers were made by drying the remaining coating liquid in Petri-dishes.

3. Preparation of microporous membrane toplayers

For the manufacturing of an ultra-thin microporous toplayer with a pore size of 1 nm or smaller, a 'so-called' polymeric type of sol-gel method was considered, in analogy with the preparation route of state of the art microporous SiO_2 membrane layers. The alkoxides of zirconium are however much more reactive towards water than Si-precursors (e.g. $Si(OC_2H_5)_4$), requiring a pretreatment with an inhibitor/stabilizer prior to performing the hydrolysis and polymerization reactions involved in the polymeric sol formation.

In a preferred preparation procedure, an alcohol amine (diethanol amine (DEA) was added to a solution of a zirconia precursor ($Zr(OC_3H_7)_4$) and an yttrium precursor ($Y(OC_4H_9)_3$) and n-propanol (n-C_3H_7OH), after which a solution of a modified precursor was obtained. A stable yttria-doped zirconia sol containing nano-particles could be obtained by hydrolysing these modified precursors with 5 mole of H_2O, in the presence of HNO_3. The essential feature of the method used here was that the alcohol amine also functions as a coating and drying controlling additive during the critical deposition process [8]. Then, a mesoporous γ-Al_2O_3 or $8Y_2O_3$-ZrO_2 membrane was dipped into the diluted sol during 15 s. Subsequently, the coating was dried in air and fired at 400 – 500°C with a heating and cooling rate of 1°C, to give a supported membrane layer on the mesoporous carrier membrane. In the preparation procedure, each dip-coating – drying – firing cycle was carried out twice.

RESULTS AND DISCUSSION

1. Preparation of membrane support

Figures 2a and 2b show an overview cross-section and detail surface micrograph of the substrate plate with a macroporous $8Y_2O_3$-ZrO_2 membrane layer, made by means of vacuum slip-casting. From the surface micrograph, a typical macroporous structure with a particle size in the range 300 – 400 nm is confirmed, which is also in accordance with the measured pore size of ~ 100 nm in Hg-porosimetry (Fisons Pascal 440). By looking at both micrographs, the substrate seems also to be adapted for deposition of a mesoporous membrane layer. After coating the macroporous layer, the surface looks sufficiently smooth and the pore size is adjusted to the typical range for further coating with a colloidal sol having a particle size < 100 nm.

(a)   (b)

Fig. 2. Micrographs of a macroporous $8Y_2O_3$-ZrO_2 membrane, vacuum-casted on a warm-pressed $8Y_2O_3$-ZrO_2 substrate plate
((a) overview cross-section micrograph (bar = 100 μm); (b) detail surface micrograph (bar = 1 μm))

2. Preparation of mesoporous intermediate membrane layers

Figures 3a and 3b show an overview and detail cross-section micrograph of a mesoporous Al_2O_3 membrane layer prepared by dip-coating with a sol containing colloidal particles with a size of ~ 30 nm. From these micrographs, a typical graded membrane structure can be clearly observed comprising a macroporous $8Y_2O_3$-ZrO_2 layer with a thickness of ~ 30 - 40 μm and a mesoporous Al_2O_3 layer with a thickness of ~ 3 - 4 μm.

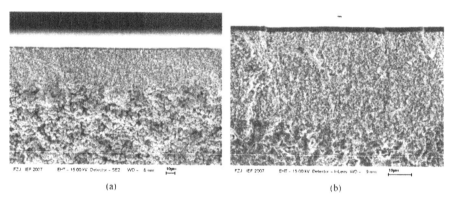

(a)   (b)

Fig. 3. Micrographs of a mesoporous γ-Al_2O_3 membrane, dip-coated on the macroporous $8Y_2O_3$-ZrO_2 membrane shown in Figure 2 (particle size in sol ~ 30 nm)
((a) overview cross-section micrograph (bar = 10 μm); (b) detail cross-section micrograph (bar = 10 μm))

By looking at the detail micrograph, it appears also that a continuous, very homogeneous and separate membrane layer has been formed and that infiltration of sol particles into the substrate macropores could be prevented using PVA as an additive during the coating step. The intermediate layer was prepared according to standard methods described for example in Ref. 9. After firing at 600°C, a comparable average pore size of ~ 4 nm and the γ-Al$_2$O$_3$ phase was observed.

Attempts to make zirconia membrane layers with a similar membrane thickness using the same preparation method and a similar particle size (~ 30 nm) failed, which is in accordance with previous findings [7]. In order to create a continuous membrane layer with a similar thickness in the micrometer range as the γ-Al$_2$O$_3$ membrane, a colloidal sol containing larger particles was tested (~ 200 nm). From the micrograph given in Figure 4a, a similar graded structure can be observed as shown in Figure 3 for the membrane with a γ-Al$_2$O$_3$ mesoporous layer. From Figure 4b, a comparable average membrane thickness of ~ 3 - 4 µm can also be estimated, for the obtained 8Y$_2$O$_3$-ZrO$_2$ layer membrane layer made by a double dip-coating – calcination procedure as described in the experimental section.

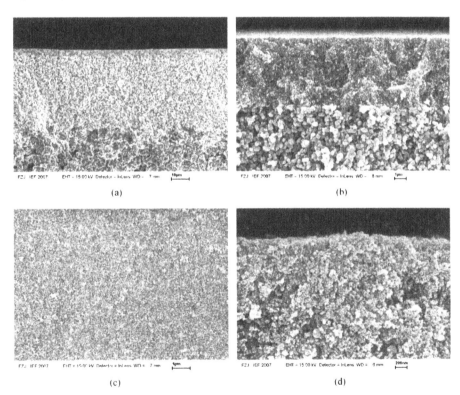

(a) (b) (c) (d)

Fig. 4. Micrographs of a mesoporous 8Y$_2$O$_3$-ZrO$_2$ membrane, dip-coated on the macroporous 8Y$_2$O$_3$-ZrO$_2$ membrane shown in Figure 2 (particle size in sol ~ 200 nm)
((a) overview cross-section micrograph (bar ~ 10 µm); (b) detail cross-section micrograph (bar ~ 1 µm); (c) detail surface micrograph (bar = 1 µm); (d) detail cross-section micrograph at large magnification (bar = 200 nm))

By comparing a detail surface image (Figure 4c) with the surface image of the supporting vacuum-casted $8Y_2O_3$-ZrO_2 macroporous layer (Figure 2b), it appears also that a $8Y_2O_3$-ZrO_2 membrane layer with a much finer (meso) porous structure was obtained. In further coating experiments, this membrane layer appeared however unsuitable as a substrate for further deposition of an ultra-thin membrane layer from a sol containing nano-particles. Apparently, the pore size of the membrane layer (diameter ~ 20 nm) was not sufficiently adapted to the particles in the nano-sol (size ~ 6 nm) in order to prevent particle infiltration. A second undesirable property which prevents the formation of a continuous ultra-thin layer could also be the significant roughness of the surface of the mesoporous membrane layer, when looking at a scale which corresponds to the thickness of ultra-thin layers (Figure 4d). This roughness can be transferred to a coating with a typical thickness in the range 50 - 100 nm, which can be an important stress factor and hinder the formation of a continuous layer.

After analysis of the previous results, the approach has been to prepare a coating liquid which gives a continuous mesoporous membrane layer on the macroporous substrate and produces at the same time a mesoporous material with a smaller pore size adapted to the size of the particles in the nano-sol. The best results were obtained using a colloidal sol with an average particle size of ~ 60 – 70 nm (Figure 5).

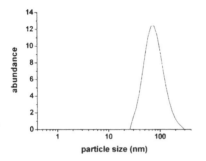
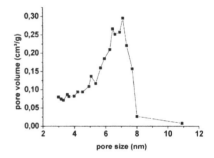

Fig. 5. Particle size distribution of $8Y_2O_3$-ZrO_2 sol used for making a mesoporous membrane layer

Fig. 6. Pore size distribution of $8Y_2O_3$-ZrO_2 mesoporous membrane material (firing 500°C)

Using such a sol, rather thin membrane layers with an average thickness of ~ 0.5 μm were obtained as shown in the micrograph given in Figure 7a, but the finer sol particles gave a clearly improved membrane layer with a decreased surface roughness and a smaller pore size. Pore analysis of the toplayer material with N_2-adsorption/desorption measurements indicated a mesoporous structure (type IV isotherm) with a BET surface area of ~ 100 m²/g and a pore size maximum of ~ 6 - 7 nm (Figure 6). The scattering in the pore size distribution can be explained by the sample preparation. After each series of coatings, the remaining sol was dried and collected. For accurate measurement, a relatively large amount of calcined powder was required and therefore dried sol from several coating experiments was used for one measurement. The phase composition of the material was characterized as single-phase cubic-ZrO_2 (Figure 10a), in accordance with data given in literature for yttria-doped zirconia powders with a similar percentage of doping.

In order to improve the thickness and quality of the mesoporous $8Y_2O_3$-ZrO_2 membrane layer, a second dip-coating step was included. As shown in Figure 7b, a membrane layer with an average thickness of ca. 1 μm could be produced in this way. In Figures 7c and 7d, images taken in the back-scattering mode of the same membrane are shown. In this mode of operation, an improved contrast can be observed between materials or membrane layers with a different porosity and a different composition (e.g. Al_2O_3, ZrO_2). In the overview back-scattering micrograph, the graded structure of the obtained multilayer membrane can be observed, with successively the substrate material, a macroporous 8Y$_2$O$_3$-ZrO$_2$ layer made from a suspension and a mesoporous 8Y$_2$O$_3$-ZrO$_2$ membrane layer made from the described colloidal sol (Figure 7c). The pore size of these layers measures 1 μm (substrate), 100 nm (macroporous layer) and 6 nm (mesoporous layer). By looking at the detail image of this membrane given in Figure 7d, the separation line between the successive membrane layers is also clearly visible and it is confirmed that the thickness of a single layer measures ~ 0.5 μm.

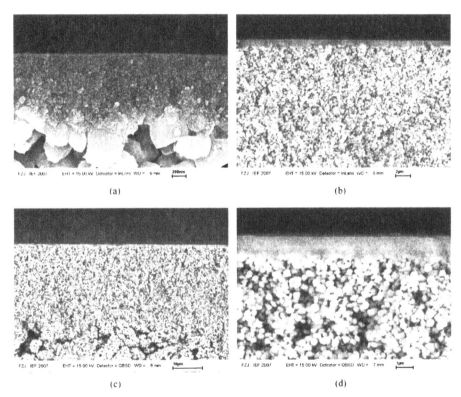

Fig. 7. Micrographs of a mesoporous 8Y$_2$O$_3$-ZrO$_2$ membrane, dip-coated on a macroporous 8Y$_2$O$_3$-ZrO$_2$ membrane layer made by screen-printing (particle size in sol ~ 65 nm)
((a) detail cross-section micrograph of mono-layer membrane (bar – 200 nm); (b) cross-section micrograph of double-layer membrane (bar ~ 2 μm); (c) overview cross-section micrograph of double-layer membrane in back-scattering mode (bar = 10 μm); (d) detail cross-section micrograph of double-layer membrane in back-scattering mode (bar ~ 1 μm))

3. Preparation of microporous membrane toplayers

For the formation of the membrane toplayer, sols with a similar particle size as standard polymeric silica sols were aimed at. Figure 8 shows that this could be obtained according to the followed sol synthesis route with an alcohol amine as precursor inhibitor for the reaction/polymerization of the precursor compounds. All of the prepared sols were stable, precipitate-free and showed an average particle size in the range 5 – 10 nm and a narrow particle size distribution. Furthermore, Figure 8 shows that according to the proposed preparation route, sols having different molar ratios can be formed, for making membrane materials with different Y_2O_3:ZrO_2 ratios (e.g. 3:97, 8:92).

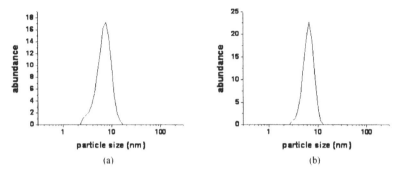

(a)   (b)

Fig. 8. Particle size distribution of yttria-zirconia sols used for the development of microporous Y_2O_3-ZrO_2 membrane layers ((a) $3Y_2O_3$-ZrO_2; ((b) $8Y_2O_3$-ZrO_2)

In dip-coating experiments, membrane formation appeared however not evident. The first coating experiments confirmed also the major importance of the properties of the mesoporous layer, which functions as carrier for the nano-structured toplayer. In literature, ultra-thin microporous SiO_2 membrane layers are typically deposited on a mesoporous γ-Al_2O_3 membrane. In our experiments, this appeared also by far the most convenient substrate to obtain a crack-free and homogeneous ultra-thin Y_2O_3-ZrO_2 membrane layer. Figure 9 shows micrographs of a $3Y_2O_3$-ZrO_2 and $8Y_2O_3$-ZrO_2 toplayer, coated on a mesoporous γ-Al_2O_3 interlayer and fired at 500°C. In micrograph 9a and 9b, a typical multilayer structure can be observed with a nano-structured Y_2O_3-ZrO_2 membrane layer as toplayer. From the detail micrograph shown in Figure 9c, it appears also that a very uniform and separate toplayer was obtained and that infiltration of sol nano-particles into the pores of the interlayer did not occur. From this micrograph, an average thickness of ~ 50 - 100 nm can be estimated for the ultra-thin toplayer. Micrograph 9d gives on overview of the same membrane in the back-scattering mode. Successively, the following layers are present in this image: a macroporous $8Y_2O_3$-ZrO_2 layer made from a suspension with a pore size of ~ 100 nm, a mesporous γ-Al_2O_3 membrane layer made from a colloidal sol with a pore size of ~ 4 nm and a nano-structured $3Y_2O_3$-ZrO_2 membrane layer made from a sol with nano-particles.

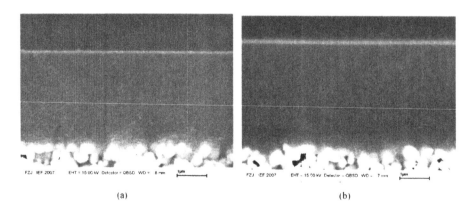

(a)   (b)

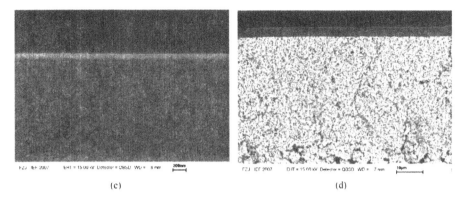

(c)   (d)

Fig. 9. Micrographs of a microporous 3Y$_2$O$_3$-ZrO$_2$ and 8Y$_2$O$_3$-ZrO$_2$ membrane, dip-coated on the mesoporous γ-Al$_2$O$_3$ membrane shown in Figure 3 (particle size in sol ~ 6 nm)
((a) cross-section micrograph of 3Y$_2$O$_3$-ZrO$_2$ membrane (bar ¯ 1 μm, back-scattering mode); (b) 8Y$_2$O$_3$-ZrO$_2$ membrane (bar ‒ 1 μm, back-scattering mode); (c) detail cross-section micrograph of 3Y$_2$O$_3$-ZrO$_2$ membrane (bar ‒ 200 nm, back-scattering mode); (d) overview micrograph (bar ‒ 10 μm, back-scattering mode))

During the coating experiments, differences between the deposition of a 3Y$_2$O$_3$-ZrO$_2$ or a 8Y$_2$O$_3$-ZrO$_2$ membrane layer were not experienced. In this work, 8Y$_2$O$_3$-ZrO$_2$ was selected as material for further studies, since the previously developed mesoporous membrane layer (shown in Figure 7) also consists of the same material. Figure 10b shows that the cubic polymorph of zirconia was also found for the membrane material made from the nano-structured sol. Pore analysis of the calcined material (firing at 450°C) indicated a microporous structure (type I isotherm), with an average pore size of ~ 1 nm (Figure 11).

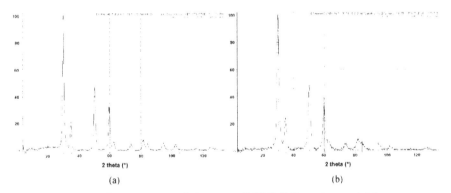

2 theta (°)   2 theta (°)
(a)   (b)

Fig. 10. XRD pattern of mesoporous (a) and microporous (b) 8Y$_2$O$_3$-ZrO$_2$ membrane material

With the obtained results, it has been shown that a crack-free microporous zirconia membrane layer with a pore size of ~ 1 nm can be produced from the nano-particle sol. Unfortunately, the very smooth supporting γ-Al$_2$O$_3$ mesoporous layer – which is also typically applied in current microporous multilayer membranes – has only a limited chemical and (hydro)thermal stability. Therefore, attempts were made to deposit a similar toplayer on the previously described membrane with a mesoporous 8Y$_2$O$_3$-ZrO$_2$ layer. As already mentioned, coating of an ultra-thin toplayer on a mesoporous zirconia membrane layer was experienced to be much more difficult, but after optimizing the coating procedure of the mesoporous intermediate layer and the composition of the nano-particle sol, a continuous ultra-thin $8Y_2O_3$-ZrO_2 nano-structured membrane layer could be obtained (Figure 12a). In the overview back-scattering micrograph in Figure 12b of the same membrane, two mesoporous $8Y_2O_3$-ZrO_2 membrane layers made from a colloidal sol and an ultra-thin microporous $8Y_2O_3$-ZrO_2 membrane layer made from the nano-structured sol can be clearly distinguished.

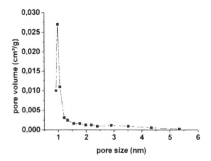

Fig. 11. Pore size distribution of $8Y_2O_3$-ZrO_2 microporous membrane material (firing 500°C)

The images shown in Figure 12 were however obtained on small substrates, which were extensively polished with diamond suspensions. Of course, such a procedure is only applicable for the development of membranes for use at lab-scale. The final step in the membrane development was then to transfer the same relatively thin mesoporous and microporous membrane coatings to the $8Y_2O_3$-ZrO_2 substrate, previously applied for the formation of multilayer membranes with a rather thick mesoporous γ-Al_2O_3 intermediate membrane layer and an ultra-thin microporous $8Y_2O_3$-ZrO_2 toplayer as shown in Figure 9.

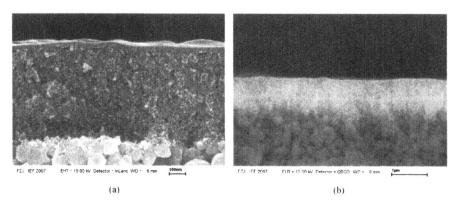

(a)   (b)

Fig. 12. Micrographs of a microporous $8Y_2O_3$-ZrO_2 membrane, dip-coated on a mesoporous $8Y_2O_3$-ZrO_2 membrane (particle size in sol ~ 6 nm, polished substrate type)
((a) cross-section micrograph (bar = 200 nm); (b) cross-section micrograph in back-scattering mode (bar = 1 μm))

Figure 13 shows a multilayer $8Y_2O_3$-ZrO_2 membrane made with the same $8Y_2O_3$-ZrO_2 substrate. In order to obtain more information on the influence of the rougher unpolished surface of the substrate on the formation of a coherent microporous membrane layer, the cross-section images were taken with an inclination of 5%. The micrograph shows that a continuous coating has been formed over the wavy surface of the mesoporous 8Y$_2$O$_3$-ZrO$_2$ layer (Figure 13a). Normally, sol-gel deposition is done on very flat and smooth surfaces, such as silicon wafers or glass plates. In this work, it is confirmed that a porous substrate, which shows also a significant surface roughness, allows the formation of a continuous ultra-thin membrane layer. The creation of a continuous membrane layer is also evident from surface micrographs made before and after the coating procedure (Figures 13b-d). Figure 13b shows a detail micrograph of the membrane surface prior to performing the dip-coating procedure with a nano-particle sol (surface of the mesoporous 8Y$_2$O$_3$-ZrO$_2$ intermediate membrane layer). Figure 13c shows the surface after dip-coating and firing at 500°C. By comparing these pictures, it appears clearly that a membrane layer with a much finer pore structure is formed on the mesoporous intermediate layer. In addition, the overview micrograph shown in Figure 13d indicates that a crack-free membrane layer was obtained.

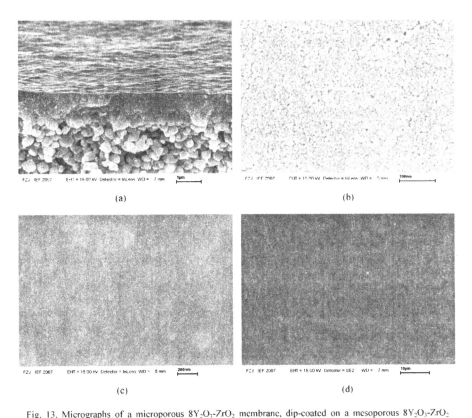

Fig. 13. Micrographs of a microporous 8Y$_2$O$_3$-ZrO$_2$ membrane, dip-coated on a mesoporous 8Y$_2$O$_3$-ZrO$_2$ membrane as shown in Figure 7 (particle size in sol ~ 6 nm)
((a) cross-section micrograph (bar = 1 μm, inclination 5°); (b) detail surface micrograph of 8Y$_2$O$_3$-ZrO$_2$ mesoporous membrane (bar = 200 nm); (c) detail surface micrograph of 8Y$_2$O$_3$-ZrO$_2$ microporous membrane (bar = 200 nm); (d) overview surface micrograph of 8Y$_2$O$_3$-ZrO$_2$ microporous membrane (bar = 10 μm))

CONCLUSION

In the present work, a series of novel porous ceramic membranes with functional toplayers made of Y$_2$O$_3$-doped ZrO$_2$ have been obtained, using classical ceramic membrane processing procedures (suspension coating, sol-gel). These membranes show a typical graded membrane structure in analogy with current ceramic gas separation membranes and the application of zirconia based materials represents a significant improvement in terms of chemical and thermal stability over current silica based materials.

For applications requiring a fine microporous structure, ultra-thin nano-structured 8Y$_2$O$_3$-ZrO$_2$ membrane layers with a thickness of ~ 50 - 100 nm were made from sols containing particles with a size of ~ 6 nm. The average pore size of the nano-structured membrane material measured ~ 1 nm, which approaches the pore size of current gas selective membranes (~ 0.4 nm). Further studies are currently devoted to tune the pore size of the novel membranes to the aimed ultra-microporous region (pore size < 0.5 nm), in order to obtain a functional separation membrane with an optimal pore size, an excellent thermal and chemical stability.

REFERENCES

[1] P. Puhlfürß, A. Voigt, R. Weber and M. Morbé, Microporous TiO$_2$ membranes with a cut-off <500 Da, J. Membr. Sci. 174 (2000) 123-133

[2] H.M. van Veen, Y.C. van Delft, C.W.R. Engelen and P.P.A.C. Pex, Dewatering of organics by pervaporation with silica membranes, Sep. Pur. Techn. 22-23 (2001) 361-366

[3] T.A. Peters, J. Fontalvo, M.A.G. Vorstman, N.E. Benes, R.A. van Dam, Z.A.E.P. Vroon, E.L.J. van Soest-Vercammen and J.T.F. Keurentjes Hollow fibre microporous silica membranes for gas separation and pervaporation: Synthesis, performance and stability, J. Membr. Sci. 248 (2005) 73-80

[4] M. Asaeda, J. Yang and Y. Sakou, Porous Silica-Zirconia (50%) Membranes for Pervaporation of iso-Propyl Alcohol (IPA)/Water Mixtures, J. Chem. Eng. Japan 35 (2002) 365-371

[5] R. Vacassy, C. Guizard, V. Thoraval and L. Cot, Synthesis and characterisation of microporous zirconia powders. Application in nanofiltration characteristics, J. Membrane Sci. 132 (1997) 109-118

[6] B.E. Yoldas, Preparation of glasses and ceramics from metal-organic compounds, J. Mat. Sci. 12 (1977) 1203-1208

[7] T. Van Gestel, H. Kruidhof, D. Blank and H.J.M. Bouwmeester, Development of microporous ZrO$_2$ and TiO$_2$ toplayers for nanofiltration (NF) and pervaporation (PV) : Part 1. Preparation and characterization of a corrosion resistant ZrO$_2$ NF toplayer with a MWCO < 300, J. Membr. Sci. 284 (2006) 128-136

[8] T. Van Gestel, D. Sebold, H. Kruidhof, H.J.M. Bouwmeester, ZrO$_2$ and TiO$_2$ membranes for nanofiltration and pervaporation: Part 2. Development of ZrO$_2$ and TiO$_2$ toplayers for pervaporation, J. Membr. Sci., in press

[9] V.T. Zaspalis, W. Van Praag, K. Keizer, J.R.H. Ross and A.J. Burggraaf, Synthesis and characterization of primary alumina, titania and binary membranes, J. Mat. Sci. 27 (1992) 1023-1035

ELECTRICAL, MECHANICAL, AND THERMAL PROPERTIES OF MULTIWALLED CARBON
NANOTUBE REINFORCED ALUMINA COMPOSITES

Kaleem Ahmad[a] and Wei Pan

State Key Lab of New Ceramics and Fine Processing, Department of Materials Science and
Engineering, Tsinghua University, Beijing 100084, People's Republic of China

ABSTRACT

The outstanding electrical, mechanical, and thermal properties of carbon nanotube (CNT) have
prompted the research for development of advanced engineering composites with improved properties.
It is based on the notion that by combining CNT with ceramics, some of the attractive properties of
CNT may be transferred to the resulting composites. Most of the studies were focused on improving
the fracture toughness of alumina by using CNT. In this work, we have investigated multifunctional
properties of multiwalled carbon nanotube (MWNT)/alumina composites to make the most of the
exceptional characteristic of MWNT in the resulting composite. Results have shown that the low
fractions of MWNT can be used successfully to convert electrically insulating alumina into electrically
conducting composites through percolating network of MWNT at the grain boundaries. In case of
mechanical properties, the fiber toughening by MWNT was found main mechanism for fracture
toughness improvement. The thermal conductivity and thermal diffusivity of the composites decrease
with MWNT contents, suggests that interfacial thermal barrier between MWNT and alumina play a
key role in determining these properties. Low volume fraction of MWNT can be used successfully to
concurrently improve electrical and mechanical properties of alumina without deteriorating other
intrinsic properties. This kind of multifunctional alumina composites with improved electrical and
mechanical properties but with reduced thermal properties can be employed for a wide range of
potential applications as structural and functional ceramics.

INTRODUCTION

In the last several years, carbon nanotubes (CNTs) have attracted much attention due to their
outstanding electrical, mechanical, and thermal properties.[1] Novel experiment suggests that tensile
strength and Young's modulus of outer layer of individual multiwall carbon nanotube (MWNT) are
11-63 GPa, and 270-950 GPa, respectively.[2] Further, experimental observations at room temperature
on individual MWNT show high thermal conductivity value of more than 3000 W/Km. In addition, the
electrical properties demonstrate a multichannel quasiballistic conducting behavior attributed to
multiple wall and large diameter with high electrical conductivity $\approx 1.85 \times 10^3$ S/cm and huge current
density $\approx 10^7$ A/cm^2 values along the long axis.[3-5] The cost effective mass production, high aspect ratio,
nanosize, and low density make them the prime candidate for development of advanced engineering
composite materials. Alumina as structural ceramics have potential or already used applications in
various fields such as, armor systems, wear resistance products, electronic substrates, cutting tools,
automotive parts, turbine hot section components, power generator components, furnace elements and
components.[6-8] However, the present alumina based ceramics have low performance limited by their
fracture toughness. Furthermore, alumina as functional ceramics can be used, where higher electrical
conductivity is required, such as, heating elements, electrical igniters, antistatic, and electromagnetic
shielding effectiveness of electronic components.[9, 10] No attempts have been made to take the full
advantage of the unique properties of MWNTs in alumina based composites. Most of the studies have
used single wall carbon nanotubes (SWNTs) due to their perfect structure[11] to improve the mechanical
properties specially, the fracture toughness of alumina.[11-13] On the other hand Xia et al.[14, 15] showed
that in imperfect MWNTs the inner wall with fracture ends have pullout forces $\approx 3\text{-}4$ times larger than

those with caped end so imperfect MWNTs may provide more effective load transfer from outer to inner walls and useful to enhance composites strength and toughness. Zhan et al.[11] claimed three times improvements in the fracture toughness over monolithic alumina with 10 volume (vol) % of SWNTs. On the contrary Wang and coworkers [13] claimed that CNTs do not improve the fracture toughness of alumina except some increase in contact damage resistance and results of Zhan[11] were over estimated using indentation method.[12, 16] Another study by Xia and colleagues reported toughening mechanisms on specially designed highly ordered parallel array of MWNTs on amorphous nanoporous thin alumina matrix (20 and 90 μm thickness).[15] However, the overall composite toughness was difficult to measure due to complex nature of residual stresses and crack bridging.[13, 15] Only a few researchers have investigated CNTs for concurrent use in electrically functional and structural ceramics, especially MWNTs/alumina composites remained a less focused area and needs to be investigated. In the present study the multifunctional properties of alumina composites reinforced by MWNTs were investigated to achieve the most of the exceptional characteristics of MWNTs.

2. EXPERIMENTAL

MWNTs were dispersed carefully to ensure maximum homogeneity of the composites using the process described by Zhan et al.[11, 17] In brief, the starting powder alumina (99.9% in purity, Japanese) was ultrasonically mixed with different (0.5, 3.0 and 5.0) weight percentage (wt %) of MWNT (Green Chemical Reaction Engineering and Technology, China) in ethanol. MWNTs used in this work were prepared by the catalytic decomposition of propylene on Fe/Al_2O_3 catalyst and having outer diameter less than 20 nm and few micrometers in length. For further mixing ball milling was performed for 24h. After drying, powder mixtures were spark plasma sintered (Dr. Sinter 1050, Sumitomo Coal Mining Co., Japan) at 1400 °C in vacuum under a pressure of 50 MPa in 20 and 12.5 mm inner diameter cylindrical graphite molds. CNTs have been reported to be stable up to 1550 °C using spark plasma sintered.[13] so any damage or change in the structure of CNTs at 1400 °C is not expected to occur. The bulk densities of sintered samples were measured by the Archimedes's method. The elastic modules of samples were measured by the ultrasonic method (Ultrasonic Pulser/Receiver Model 5900 PR, Panametrics, Waltham, MA). The fracture toughness was measured by direct toughness measurement technique i.e. single edge notch beam method on rectangular bars with size of about 3 mm ×2mm × 15mm containing a pre-notch of length ≤1.5mm and width ≈0.25mm at cross head speed of 0.05 mm/min. The test was performed at room temperature using a universal testing machine (Shimadzu Servo Pulser EHF- EG50KNT-10L, Japan). Three bars were tested for each material. The hardness was measured by Vickers indentation method with a load of 5kg. Cracks on polished surfaces were produced to investigate the possible toughening mechanisms. The scanning electron microscopes HITACHI –S-4500 was used to study the morphology of the fracture surface and to identify the toughening mechanism of the composites. Silver paste was painted on both sides of disc shaped samples of diameter ≈12.5 mm and room temperature dc electrical conductivity was measured by two point probe method. The thermal diffusivity was measured along the thickness of the disc shaped samples ≈12.5mm in diameter using the laser flash technique (NETZSCH Laser Flash Apparatus LFA 427) in the temperature range from ambient temperature to 500 °C in argon atmosphere.

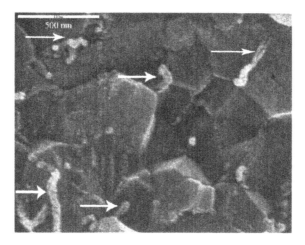

Fig. 1 Uniform distribution of multiwalled carbon nanotubes for 3 wt % of MWNT/alumina composites

3. RESULT AND DISCUSSION

Mechanical Properties

Multiwalled carbon nanotubes are uniformly distributed in alumina matrix at intergranular positions as shown in Figure 1. The average grain size of alumina is around 350nm. Mechanical properties and relative density of alumina nanocomposites as a function of MWNTs wt % are shown in Table. 1. The relative densities of 0, 0.5, 3.0, and 5.0 wt % of MWNTs reinforced alumina composite reached at 99.5, 98.5%, 95.1%, and 92.0%, respectively. The relative density of the composites decreases with increase of MWNT contents. As the wt % increase the probability of agglomeration of MWNTs also increases that may cause decrease in the density.

Table 1. Processing conditions and resultant properties of multiwalled carbon nanotube reinforced alumina nanocomposites consolidated by spark plasma sintering.

Materials	Processing conditions	Relative Density (%TD)	Fracture Toughness (MPam$^{1/2}$)	Hardness (GPa)	Young's Modulus (GPa)
Al$_2$O$_3$	SPS 1400 °C 3 min	99.5	3.3	17	400.48
0.5 wt% MWNT- Al$_2$O$_3$	SPS 1400 °C 3 min	98.5	3.78	19	394.21
3.0 wt% MWNT- Al$_2$O$_3$	SPS 1400 °C 3 min	95.1	5.55	15	313.03
5.0 wt% MWNT- Al$_2$O$_3$	SPS 1400 °C 3 min	92.0	4.90	11	277.39

Some clustering is observed at 5 wt % of MWNT/alumina composites that reduced the density. Ning et al. also observed decrease in density with the increase of MWNTs contents in SiO_2 matrix and suggest that MWNT act as a kind of solid impurity which prevent the flowing of the matrix.[18]

The hardness of alumina nanocomposites shows dependency on the density (Table. 1). The hardness increases for 0.5 wt % of MWNTs and then it decreases with decrease in relative density. The hardness seems to strongly dependent on the density which decreased moderately at 3 wt % and further decreased drastically at 5 wt % of MWNTs/alumina composites due to decrease in the density. Zhan et al.[11] showed a sharp decrease in hardness with the increase of fracture toughness, while some other authors have reported an increase in hardness, but in their work there is no consistent relationship between hardness and fracture toughness.[19, 20]

The fracture toughness of the composites increases with increase of MWNTs wt %, while at 5 wt % it decreases slightly. The fracture toughness at 5 wt % decreased due to agglomeration of MWNTs. The agglomeration of MWNTs reduces the density and consequently less improvement in the mechanical properties specially the fracture toughness. The significant toughness enhancement is dominantly attributed to MWNTs through fiber toughening mechanisms. As the wt % increases the fiber toughening is also increases which results in improvement of fracture toughness. The SEM observation of fiber toughening (crack bridging, MWNT pull outs) by the MWNTs provides direct evidence for improvement in the toughness of the composites (Fig. 2). The crack bridging by MWNTs try to restrict the crack opening size and reduces the driving force for crack propagation. The pull outs indicate that MWNTs bear large stresses by sharing the portion of the load and debonding occurs at the atomic scale. In addition, the work requires pulling MWNTs out against residual sliding friction at the interface imparts significant fracture toughness improvement to the alumina matrix. Zhan et al.[11] and Fan et al.[21] reported significant improvement in the fracture toughness of CNTs/alumina composites however, in their studies the enhanced toughening was not supported by any key evidence of potential toughening mechanisms. In the present study, the nanocomposites have shown an improvement $\approx$ 68.2 % in the fracture toughness for 3 wt % MWNTs/alumina composites.

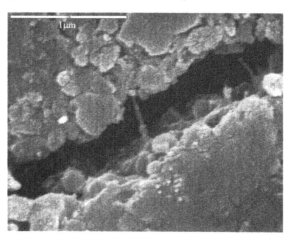

Fig. 2 Fiber toughening by multiwalled carbon nanotubes in 3 wt % MWNT/alumina composite

Furthermore, the observed significant enhancement in the fracture toughness has been substantiated by direct evidence of fiber toughening mechanism through a nanoscale uniform dispersion of highly disordered MWNTs. Transmission electron microscopy has already shown a sharp interface between graphen wall of CNT and alumina, which suggests a good bonding between them.[22] There are possibly two factors complementing each other in toughening the alumina matrix. Firstly, the good bonding between MWNTs and alumina as suggested earlier[11, 22] and secondly, the presence of a network structure of MWNTs at intergranular positions as reported in several studies[17, 23] may result in strengthening and toughening of the nanocomposites by MWNTs. The fiber toughening by MWNTs may be predominantly attributed to the improvement in the toughness.

The Young's modules of alumina and composites measured by ultrasonic spectroscopy are shown in Table 1. The Young's modules of the composites decrease with increase of MWNT contents. The Young's modules of the composites depend on the constituent materials and porosity. The Young's modules of individual MWNT in the lower range ($\geq$ 270 GPa) range[2] is almost comparable to Young's modules of alumina (400 GPa). The value of Young's modules of the composites may decreases due to the low value of Young's modules of MWNT ropes and porosity in the composites. It is generally reported that porosity reduces Young's modulus because pores act as a second phase of zero modulus.[24]

Thermal Properties

The temperature dependence of experimental thermal diffusivity and calculated thermal conductivity are presented in Fig. 3 and 4, respectively. The thermal diffusivities of the composites decrease with increasing temperature and follow a power law. This can be explained by the fact that the thermal conduction in composites is dominated by phonons. The thermal conductivity (λ_{TC}) of the samples was calculated from the thermal diffusivity coefficient κ, bulk density ρ, and specific heat C_p by using the following standard equation.

$$\lambda_{TC} = \kappa \, \rho \, C_p. \qquad (1)$$

The specific heat can be approximated using the rule of mixture. In the preset study, the specific heat C_p of the composites was calculated using the rule of mixture and the effect of temperature on C_p was also taken into account.

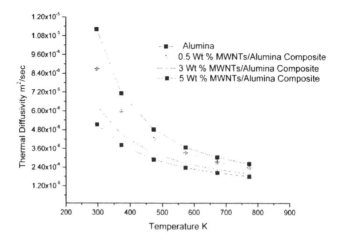

Fig. 3 Temperature dependence of thermal diffusivity of MWNT/alumina composites

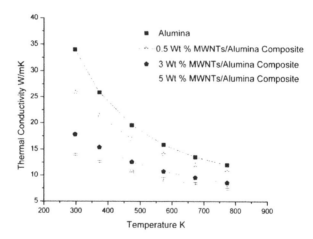

Fig. 4 Temperature dependence of calculated thermal conductivity of MWNT/alumina composites

The specific heat values for the MWNTs, and alumina were obtained from the thermodynamic database[25] and the specific heat of graphite was used for MWNTs due to structural similarity of both carbon materials as used in other studies.[26, 27] It is surprising to find that the thermal conductivity of MWNTs /alumina composites decreases with increase of MWNTs contents. The effective thermal conductivity of a composite is related to the conductivities of the matrix and inclusions. In the present study, firstly, the decrease in thermal conductivity by addition of MWNT may be attributed to low thermal conductivity of MWNT ropes, since the effective resistance between two nanotubes severely reduces the thermal conductivity of the ropes in comparison with individual MWNT. Secondly, the interactions between nanotubes and the surrounding matrix provides sufficient scattering of phonons to reduce effective conductivity of the composites. It has been reported that the thermal conductivity is controlled by interface thermal conductance,[28] which can significantly alter the effective thermal conductivity of CNTs reinforced composites. Due to interfacial thermal resistance, the decrease in thermal conductivity by addition of CNTs have been reported in ceramics by several authors.[29, 30] The effect of low thermal conductivity of carbon nanotube ropes, interface thermal resistance or Kapitza resistance between alumina matrix and MWNTs due to scattering of phonon may be the possible causes of decreases in thermal conductivities in carbon nanotube reinforced alumina composites.

Electrical Conductivity

Figure 5 shows the room temperature DC electrical conductivity of alumina composites as a function of MWNT wt %. The conductivity increases dramatically at 0.5 wt% of MWNT and then it levels off for further addition of MWNT and shows typical conductor insulator transition. Percolation theory defines an insulator-conductor transition and a corresponding threshold of the conductive filler concentration P_{MWNT}, through the following equation.

$$\sigma = \sigma_0 (P - P_{MWNT}) \quad \text{for } P > P_{MWNT} \qquad (2)$$

Where σ_0 is a constant, p the weight fraction of nanotubes, and t the critical exponent. The data are fitted to the scaling law, and the best-fitted values are for $P_{MWNT} = 0.45 \pm 0.05$ wt % and t=1.78±0.02. The low value of percolation threshold $P_{MWNT} = 0.45$ wt % is attributed to high aspect ration of MWNT. The switching from electrically insulating alumina to electrically conducting alumina composites is due to addition of MWNTs, which provides a spanning network of CNTs traveling along the grain boundaries through the alumina matrix.

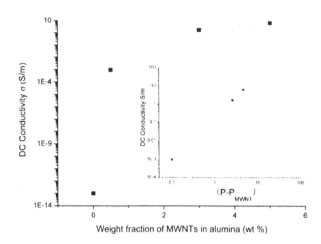

Fig. 5 Electrical conductivity of MWNT/alumina composites as a function of the MWNTs contents

The key difference between thermal transport and electrical transport is the value of the conductivity ratio between matrix and filler. For thermal transport, even for very conductive CNTs, ratio between CNTs and matrix is about 10^3-10^4, while for electrical transport the ratio of conductivities can be of the order of 10^{12}-10^{16}. Therefore, in case of electrical conductivity with such high ratio, the effective path is through carbon nanotubes, while in case of thermal conductivity, the dominant channel of heat energy flow always involve the matrix.[31]

4. SUMMARY

In summary, alumina composites reinforced by 0.5, 3.0 and 5.0 wt % of MWNT were fabricated by spark plasma sintering. The fiber toughening mechanisms by MWNTs have been observed for improvement in the fracture toughness of alumina. Further, MWNTs have been used successfully to convert electrically insulating alumina into electrically conducting composites and follow a power law of percolation with a low value of percolation threshold (0.45 wt %). The decrease in the thermal conductivity with increase of MWNTs may be attributed to the dominant effect of interface thermal resistance between alumina and MWNTs.

ACKNOWLEDGMENT

The authors acknowledge the financial support from the Higher Education Commission of Pakistan (HEC) and National Natural Science Foundation of China (Grant No. 50232020 and 50572042).

REFERENCES

[1]P. J. F. Harris, Carbon Nanotube Composites, *Int. Mater. Rev.*, **49**, 31-43 (2004).

[2]M. F. Yu, O. Lourie, M. J. Dyer, K. Moloni, T. F. Kelly, and R. S. Ruoff, Strength and Breaking Mechanism of Multiwalled Carbon Nanotubes under Tensile Load, *Science*, **287**, 637-40 (2000).

[3]H. J. Li, W. G. Lu, J. J. Li, X. D. Bai, and C. Z. Gu, Multichannel Ballistic Transport in Multiwall Carbon Nanotubes, *Phys. Rev. Lett.*, **95**, (2005).

[4]P. Kim, L. Shi, A. Majumdar, and P. L. McEuen, Thermal Transport Measurements of Individual Multiwalled Nanotubes, *Phys. Rev. Lett.*, **8721**, (2001).

[5]Y. Ando, X. Zhao, H. Shimoyama, G. Sakai, and K. Kaneto, Physical Properties of Multiwalled Carbon Nanotubes, *Int. J. Inorg. Mater.*, **1**, 77-82 (1999).

[6]A. Krell, P. Blank, L. M. Berger, and V. Richter, Alumina Tools for Machining Chilled Cast Iron, Hardened Steel, *Am. Ceram. Soc. Bull.*, **78**, 65-73 (1999).

[7]E. Medvedovski, Alumina-Mullite Ceramics for Structural Applications, *Ceram. Int.*, **32**, 369-75 (2006).

[8]J. Dusza, and P. Sajgalik, Si3n4 and Al2o3 Based Ceramic Nanocomposites, *Int. J. Mater. Prod. Technol.*, **23**, 91-120 (2005).

[9]J. M. Thomassin, X. Lou, C. Pagnoulle, A. Saib, L. Bednarz, I. Huynen, R. Jerome, and C. Detrembleur, Multiwalled Carbon Nanotube/Poly(Epsilon-Caprolactone) Nanocomposites with Exceptional Electromagnetic Interference Shielding Properties, *J. Phys. Chem. C*, **111**, 11186-92 (2007).

[10]W. A. Curtin, and B. W. Sheldon, Cnt-Reinforced Ceramics and Metals, *Materials Today*, **7**, 44-49 (2004).

[11]G. D. Zhan, J. D. Kuntz, J. L. Wan, and A. K. Mukherjee, Single-Wall Carbon Nanotubes as Attractive Toughening Agents in Alumina-Based Nanocomposites, *Nat. Mater.*, **2**, 38-42 (2003).

[12]J. P. Fan, D. M. Zhuang, D. Q. Zhao, G. Zhang, M. S. Wu, F. Wei, and Z. J. Fan, Toughening and Reinforcing Alumina Matrix Composite with Single-Wall Carbon Nanotubes, *Appl. Phys. Lett.*, **89**, (2006).

[13]X. T. Wang, N. P. Padture, and H. Tanaka, Contact-Damage-Resistant Ceramic/Single-Wall Carbon

Nanotubes and Ceramic/Graphite Composites, *Nat. Mater.*, **3**, 539-44 (2004).

[14]Z. Xia, and W. A. Curtin, Pullout Forces and Friction in Multiwall Carbon Nanotubes, *Phys. Rev. B*, **69**, (2004).

[15]Z. Xia, L. Riester, W. A. Curtin, H. Li, B. W. Sheldon, J. Liang, B. Chang, and J. M. Xu, Direct Observation of Toughening Mechanisms in Carbon Nanotube Ceramic Matrix Composites, *Acta Mater.*, **52**, 931-44 (2004).

[16]B. W. Sheldon, and W. A. Curtin, Nanoceramic Composites: Tough to Test, *Nat. Mater.*, **3**, 505-06 (2004).

[17]G. D. Zhan, J. D. Kuntz, J. E. Garay, and A. K. Mukherjee, Electrical Properties of Nanoceramics Reinforced with Ropes of Single-Walled Carbon Nanotubes, *Appl. Phys. Lett.*, **83**, 1228-30 (2003).

[18]J. W. Ning, J. J. Zhang, Y. B. Pan, and J. K. Guo, Fabrication and Mechanical Properties of Sio2 Matrix Composites Reinforced by Carbon Nanotube, *Mater. Sci. Eng. A-Struct. Mater. Prop. Microstruct. Process.*, **357**, 392-96 (2003).

[19]S. I. Cha, K. T. Kim, K. H. Lee, C. B. Mo, and S. H. Hong, Strengthening and Toughening of Carbon Nanotube Reinforced Alumina Nanocomposite Fabricated by Molecular Level Mixing Process, *Scr. Mater.*, **53**, 793-97 (2005).

[20]C. B. Mo, S. I. Cha, K. T. Kim, K. H. Lee, and S. H. Hong, Fabrication of Carbon Nanotube Reinforced Alumina Matrix Nanocomposite by Sol-Gel Process, *Mater. Sci. Eng. A-Struct. Mater. Prop. Microstruct. Process.*, **395**, 124-28 (2005).

[21]J. P. Fan, D. Q. Zhao, M. S. Wu, Z. N. Xu, and J. Song, Preparation and Microstructure of Multi-Wall Carbon Nanotubes-Toughened Al2o3 Composite, *J. Am. Ceram. Soc.*, **89**, 750-53 (2006).

[22]R. Poyato, A. L. Vasiliev, N. P. Padture, H. Tanaka, and T. Nishimura, Aqueous Colloidal Processing of Single-Wall Carbon Nanotubes and Their Composites with Ceramics, *Nanotechnology*, **17**, 1770-77 (2006).

[23]A. L. Vasiliev, R. Poyato, and N. P. Padture, Single-Wall Carbon Nanotubes at Ceramic Grain Boundaries, *Scr. Mater.*, **56**, 461-63 (2007).

[24]G. H. Li, Z. X. Hu, L. D. Zhang, and Z. R. Zhang, Elastic Modulus of Nano-Alumina Composite, *Journal of Materials Science Letters*, **17**, 1185-86 (1998).

25

[26]H. L. Zhang, J. F. Li, B. P. Zhang, K. F. Yao, W. S. Liu, and H. Wang, Electrical and Thermal Properties of Carbon Nanotube Bulk Materials: Experimental Studies for the 328-958 K Temperature Range, *Phys. Rev. B*, **75**, (2007).

[27]C. Qin, X. Shi, S. Q. Bai, L. D. Chen, and L. J. Wang, High Temperature Electrical and Thermal Properties of the Bulk Carbon Nanotube Prepared by Sps, *Mater. Sci. Eng. A-Struct. Mater. Prop. Microstruct. Process.*, **420**, 208-11 (2006).

[28]S. T. Huxtable, D. G. Cahill, S. Shenogin, L. P. Xue, R. Ozisik, P. Barone, M. Usrey, M. S. Strano, G. Siddons, M. Shim, and P. Keblinski, Interfacial Heat Flow in Carbon Nanotube Suspensions, *Nat. Mater.*, **2**, 731-34 (2003).

[29]Q. Huang, L. Gao, Y. Q. Liu, and J. Sun, Sintering and Thermal Properties of Multiwalled Carbon Nanotube-Batio3 Composites, *Journal of Materials Chemistry*, **15**, 1995-2001 (2005).

[30]G. D. Zhan, and A. K. Mukherjee, Carbon Nanotube Reinforced Alumina-Based Ceramics with Novel Mechanical, Electrical, and Thermal Properties, *Int. J. Appl. Ceram. Technol.*, **1**, 161-71 (2004).

[31]N. Shenogina, S. Shenogin, L. Xue, and P. Keblinski, On the Lack of Thermal Percolation in Carbon Nanotube Composites, *Appl. Phys. Lett.*, **87**, 133106 (2005).

MICROSTRUCTURE AND DIELECTRIC PROPERTIES OF NANOSTRUCTURED TiO₂ CERAMICS PROCESSED BY TAPE CASTING

Sheng Chao, Vladimir Petrovsky, Fatih Dogan

Department of Materials Science and Engineering
Missouri University of Science and Technology
Rolla, MO 65409, USA

ABSTRACT

Effects of different solvent systems and dispersant on tape casting slurries of TiO2 nanopowders (~40nm) were studied by rheological measurements. It was found that Xylene/Ethanol with phosphate ester (PE) as dispersant gave the best dispersion quality which resulted in highly homogenous, dense, smooth and crack-free tapes. The presence of phosphate ester inhibited the sintering of TiO_2 powders and refined the grain size. Sintered tapes with 200-300nm grain size and relative density over 90%, had dielectric constant ~100, dielectric loss less than 1% and dielectric breakdown strength (BDS) over 1000KV/cm.

INTRODUCTION

Nanostructured titanium oxide ceramics attracted much attention over the decades due to its unique photocatalytic[1-3], mechanical[4] and dielectric properties[5]. Our previous research demonstrated that nanostructured TiO_2 bulk ceramics possess high dielectric breakdown strength (>1500KV/cm), low leakage current and low dielectric loss, which makes it ideal for high energy density capacitor applications.[6]

Tape casting is the most widely used method to prepare large area ceramic dielectric layer in capacitor industry. Although, tape casting is a well-established ceramic processing method, using nanopowders to prepare stable slurry for tape casting is still a very challenging task. Selection of appropriate dispersant and solvent is critical to obtain high quality tapes since preparation of dispersed slurries with high solids loading is a challenging task[7-9].

The objective of this study is to study the effect of phosphate ester as a dispersant and different solvent systems on the properties of TiO_2 slurries. The role of residual phosphorus on the sintering behavior of TiO_2 was studied and correlated with the dielectric properties of the sintered tapes.

EXPERIMENTAL

TiO_2 nanosized powders (Nanophase Technologies Corporation, IL, USA) with an average particle size of 40nm and specific surface area of $38m^2/g$ were used. The crystal structure of the starting powder is 20% of rutile phase and 80% of anatase phase.

Slurries with a fixed 20vol% solid loading were prepared by mixing TiO_2 powders with appropriate amount of solvent and dispersant for viscosity measurements. Phosphate ester (Emphos PS-21A, Witco, USA) was used as dispersant. And three commonly used solvent systems (Xylene/Ethanol, Toluene/Ethanol and MEK/Ethanol) (Table I) were employed in this experiment. Viscosity measurement was performed on a viscometer (VT550, Gebruder HAAKE, Germany) in the shear rate range between $100 s^{-1}$ and $500 s^{-1}$. Before viscosity measurements, ball milling was conducted for 24hs in polyethylene bottles using cylindrical ZrO_2 as milling media.

Table I. Chemicals used to prepare the slurries

Constituents		
Solvent	Xylene/Ethanol	(65/35, in volume)
	Toluene/Ethanol	(40/60, in volume)
	MEK/Ethanol	(65/35, in volume)
Dispersant	Phosphate ester	
Binder	PVB-79 (PVB)	
Plasticizer	Dioctyl phthalate (DOP)	

Slurry for tape casting (table II) was prepared by a two-stage method. Powders, solvent and dispersant were first ball milled for 24hrs. Binder and plasticizer were then added into the slurry and milled for another 24hrs. The tapes were cast on silicone coated Mylar at a thickness of 0.05mm. After drying, two layers of green tapes were laminated at 85°C and 3MPa pressure for 20 minutes. The tapes were sintered in pure oxygen atmosphere at 980°C for 6-12hrs.

Table II. Composition of the slurry used for tape casting

Composition	(vol%)
Powder	13.2
Xylene/Ethanol	74.8
PE	1.1
PVB	6.5
DOP	4.4

Dielectric constant and loss of selected samples were measured using Solartron 1260 impedance analyzer connected with a Solartron 1296 dielectric interface (Solartron analytical, Hampshire, UK). D.C. dielectric breakdown strength test was conducted using Spellman SL30 high voltage generator (Spellman high voltage electronics corporation, New York, USA). Samples for breakdown strength test were ~50 μm in thickness. Green and sintered tapes were characterized by scanning electron microscopy (SEM, Hitachi S4700, Japan).

RESULTS AND DISSCUSSION
Rheological Properties
Phosphate ester is a common dispersant used to prepare ceramic slurries stabilized by both steric hinderance mechanism and electrostatic repulsion mechanisms. As an anionic dispersant, the hydroxyl group of phosphate is ionized to form negatively charged oxygen.[10] Previous studies[7,11] showed that phosphate ester is an effective dispersant for various oxide powders especially BaTiO$_3$
In addition to the dispersant, the solvents used to prepare the slurries also play an important role in dispersion of the powders. The viscosity as a function of the shear rate is shown in Fig. 1 for slurries using three selected solvent mixtures with 20vol% solid loading and 1.2 wt% PE (as a measure to the powder weight). . It is depicted that the viscosity decreases with increasing shear rate, indicating a shear-thinning behavior. At the shear rates between 100 s^{-1} and 500 s^{-1}, the viscosity of the slurry with Xylene/Ethanol as solvent is consistently lower than that of other solvent mixtures. Hence, solvent mixture of Xylene/Ethanol is considered to be the best solvent system in this study.

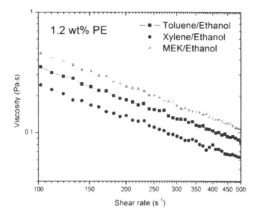

Fig.1. Viscosity vs. shear rate of slurries for various solvent systems with PE as dispersant

It is known that both solvent and dispersant will compete with each other for the absorption on the surface of the particles.[7] Polar liquids with strong hydrogen bonding capability will have strong affinity to oxide surface.[12] As a results, polar liquid is expected to be able to reduce the effectiveness of dispersant. Since ethanol is a polar, MEK weekly polar, toluene and xylene are non-polar liquids. xylene/ethanol solvent mixture (65:35) with minimum amount of polar liquids shows the lowest viscosity.

Viscosity of slurries prepared as 20vol% solid loading using xylene/ethanol as solvent with various amount of PE, was measured to determine the optimum amount of dispersant. Fig. 2 shows that the viscosity of the slurry with 1.8 wt% PE is the lowest. Thus. slurries for tape casting of titania powders were prepared using 1.8 wt% PE as a dispersant.

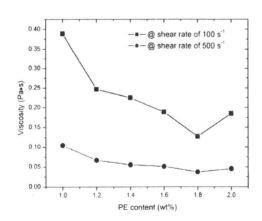

Fig. 2. Effect of the amount of phosphate ester on the viscosity of titanium oxide slurries

Sintering and Characterization of the Tapes

The microstructure of the green tape with the composition listed in Table II and a tape after binder burn-out (0.5°C/min up to 550°C) are shown in Fig. 3. It revealed that the tape was composed of TiO$_2$ nanosized powders uniformly surrounded by a layer of binder which increased the average diameter of particles to about 100nm by agglomeration. After binder burn-out, it can be seen that the average particle size decreased to the size of initial particles. The tapes were crack-free without any large pores. By measuring the weight of the tape before and after binder burn-out, the green density of the tape after binder burn-out process was calculated to be 57.4%.

Sintering of the tapes was conducted in oxygen atmosphere at 980°C between 6 and 12 hours holding time. The surface of the tapes after sintering at various holding times are revealed in Fig. 4. Residual pores present at the triple grain boundaries were observed. The relative density of the sample sintered at 980°C for 12hrs was 93.26%. However, our previous research[6] showed that for bulk samples prepared by pressing of powders without additives, a relative density over 99% could be obtained after sintering at 900°C for 2hrs. Since the green density after binder burn-out process of the tape is almost the same as that of the bulk sample after isostatic pressing (57.8%), lower sintering density of the tapes may be attributed is to the phosphate impurities introduced by phosphate ester as dispersant. It was also observed that the grain growth was suppressed for the samples prepared using PE. In order to confirm the effect of dispersant, bulk samples with 1.8 wt% PE addition were pressed and sintered. Fig. 5 shows the microstructure of the samples prepared with and without phosphate ester. It is depicted that the average grain size of the sample with PE addition sintered at 980°C for 9hrs is smaller than that of the sample without PE addition sintered at 900°C for 2hrs.

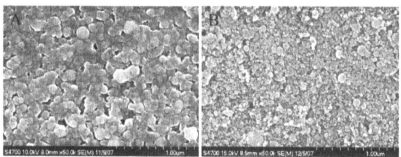

Fig. 3. Surface of green tape (A) before binder burn-out, and (B) after binder burn-out

Similar observations were made for BaTiO$_3$ prepared using phosphate ester by Caballero et. al[11] It was suggested that the rate of grain growth was lowered as a results of reduced grain boundary mobility caused by a solute drag mechanism due to the presence of phosphorus left by phosphate ester. As an effective dispersant, PE is expected to be strongly absorbed on the surface of TiO$_2$ particles, resulting in a homogeneous distribution of residual phosphorus around TiO$_2$ particles after burn-out process. As a result, the mobility of the grain boundary may be reduced due to the so called "pinning effect". Although the influence of residual phosphorus on sintering is evident, phosphorous could be not detected by energy dispersive spectroscopy (EDS) techniques applied for the sintered samples.

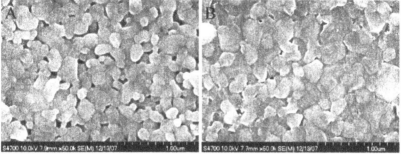

Fig. 4. Surface of the tapes after sintering at 980°C for (A) 6hrs and (B) 12hrs

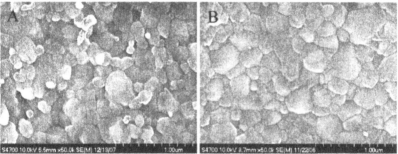

Fig. 5. Surface of sintered bulk samples (A) with PE addition at 980°C for 9hrs, and (B) without PE addition at 900°C for 2hrs

Dielectric Properties

Dielectric properties of the tapes and bulk samples are summarized in Table III. It was observed that the dielectric constant of the sintered tapes increases with the increase of soaking time while dielectric loss decreases with the increase of soaking time. This trend may be attributed to the increased sintering density. However, the breakdown strength is similar for samples sintered for 6hrs and 12hrs. Our previous studies[6] showed that breakdown strength is sensitive to both porosity and grain size. High porosity results in significant lowering of breakdown strength because it not only reduces the effective sample volume but also leads to local electric field concentration. Small grain size sample was found to exhibit higher breakdown strength than large grain size sample probably due to increased grain boundary density.

In comparison with the tape, the bulk sample shows improved dielectric properties, which may be attributed to its near full density. It should be noted that the effect of residual phosphorus on dielectric properties of titanium oxide is not well understood and requires further studies.

Table III. Dielectric properties of TiO$_2$ sintered tape and bulk ceramic[6]

Sintering conditions	Relative density (%)	Dielectric constant @ 1K Hz	Dielectric loss @ 1K Hz (%)	BDS (KV/cm)
980°C 6hrs (tape)	91.78	89.7	0.68	1039±92
980°C 12hrs (tape)	93.26	96.4	0.57	1006±230
900°C 2hrs (bulk)	99.13	138.9	0.34	1384±189

SUMMARY

Xylene/Ethanol solvent mixture and phosphate ester were found to be suitable as solvent system and dispersant for tape casting of TiO$_2$ nanosized powders. The optimum amount of the dispersant was determined to be about 1.8wt% by viscosity measurement. Residual phosphorus was found to inhibit the densification process and suppress the grain growth. Samples with dielectric constant close to 100, dielectric loss less than 1% and breakdown strength over 1000KV/cm were obtained by sintering of titania tapes with at 980°C for 6 to 12hrs.

REFERENCES
[1] F. Sayılkan, M. Asiltürk, P. Tatar, N. Kiraz, Ş. Şener, E. Arpaç and H. Sayılkan, Photocatalytic performance of Sn-doped TiO$_2$ nanostructured thin films for photocatalytic degradation of malachite green dye under UV and VIS-lights. Mater. Res. Bull., 43, 127-134, (2008).
[2] S. Zhang, Q. Yu, Z. Chen, Y. Li and Y. You, Nano-TiO$_2$ particles with increased photocatalytic activity prepared by the miniemulsion method, Mater. Lett., 61, 4839-4842, (2007).
[3] N. Venkatachalam, M. Palanichamy and V. Murugesan, Sol–gel preparation and characterization of nanosize TiO$_2$: Its photocatalytic performance, Mater. Chem. Phys., 104, 454-459, (2007).
[4] P. Ctibor, P. Boháč, M. Stranyánek and R. Čtvrtlík, Structure and mechanical properties of plasma sprayed coatings of titania and alumina, J. Euro. Ceram. Soc., 26, 3509-3514, (2006).
[5] G. Zhao, R. P. Joshi, V.K. Lakdawala, E. Schamiloglu, and H. Hjalmarson, TiO$_2$ breakdown under pulsed conditions, J. Appl. Phys., 101, 26110, (2007).
[6] S. Chao, V. Petrovsky, F. Dogan, Effects of Thermal History on the Dielectric Properties of Nanostructured Titania, Proceeding of Energy Materials and Technologies I. Materials Science and Technology (MS&T) 2007: P203-212
[7] S. Bhaskar Reddy, P. Paramanano Singh, N. Raghu, V. Kumar, Effect of type of solvent and dispersant on NANO PZT powder dispersion for tape casting slurry, J. Mater. Sci., 37, 929–934, (2002).
[8] V. Vinothini, Paramanand Singh, M. Balasubramanian, Optimization of barium titanate nanopowder slip for tape casting, J. Mater. Sci., 41, 7082–7087, (2006).
[9] K. H. Zuo, D. L. Jiang, J. X. Zhang and Q. L. Lin, Forming nanometer TiO$_2$ sheets by nonaqueous tape casting, Ceram. Int., 33, 477–481, (2007).
[10] Ging H. Hsiue, Li W. Chu and I.N. Lin, Optimized phosphate ester structure for the dispersion of nano-sized barium titanate in proper non-aqueous media, Coll. Surf. A: Physicochemical and engineering aspects, 294, 212-220, (2007).
[11] A.C. Caballero, J.F. Fernadez, C. Moure and P. Duran, Effect of Residual Phosphorus Left by Phosphate Ester on BaTiO$_3$ Ceramics, Mater. Res. Bull., 32, 221-229, (1997).

THE SIMULATION IN THE REAL CONDITIONS OF ANTIBACTERIAL ACTIVITY OF TiO$_2$ (Fe) FILMS WITH OPTIMIZED MORPHOLOGY

M. Gartner, C. Anastasescu, M. Zaharescu
Institute of Physical Chemistry "Ilie Murgulescu" of the Romanian Academy, 202 Splaiul Independentei, 060021 Bucharest, Romania

M. Enache, L. Dumitru,
Institute of Biology, 296 Splaiul Independentei, 060031 Bucharest, Romania

T.Stoica, T.F. Stoica
National Institute of Materials Physics, P.O. Box MG7, Magurele, Bucharest, Romania

C. Trapalis
Institute of Materials Science, National Center for Scientific Research "Demokritos", 15310, Athens, Greece

ABSTRACT

The aim of this study was to provide and characterize nanostructured TiO$_2$ films for application in the depollution of the contaminated water against *Escherichia coli (E. coli)*. The nanometric size of particles and the porosity of the films have been correlated with the technological parameters of the deposition process. For this aim three-layer TiO$_2$ films were deposited by sol-gel and dip-coating method on SiO$_2$/glass substrate. Iron (1 and 7%) and polyethylene glycol (PEG$_{600}$) (0.014-0.110 M) were added to the coating solutions to study their effect on the porosity of the samples and respectively on the amorphous to anatase phase transition, as well as on the antibacterial activity.

The samples were characterized using different complementary methods such as X-ray Diffraction (XRD), Spectroscopic Ellipsometry (SE), Atomic Force Microscopy (AFM), dynamics (growth and adherence) of *E. coli* development versus nanofilm composition (where the amount of bacterial inoculums used was closer to those found in wasted waters).

The obtained films have shown low roughness (values bellow 0.74 nm) and the particle size in nanometric range. The direct optical band gap values increases with the Fe content up to 2.64 eV (470 nm), leading to an active photocatalyst under visible light (in our case under natural illumination condition). The inhibitory effect towards the *E. coli* is correlated with PEG and Fe concentration and was found to be the strongest for the sample containing 0.069 M PEG.

INTRODUCTION

The use of the photocatalysts to organic compounds degradation has been intensively investigated in last years[1]. TiO$_2$ mediated photocatalytic degradation is known to mineralize a wide range of organic pollutants with faster rates than conventional treatments[2]. TiO$_2$ nanoparticles are widely used for applications such as photocatalysts, pigments, and cosmetic additives due to a series of favorable characteristics including: non-toxicity, high surface area, medium band gap, low cost, recyclability, high photoactivity, chemical and photochemical stability. The first photoelectrochemical cell for splitting water[3] based on TiO$_2$ has been reported already in '70s. The sol–gel method (liquid-phase synthesis) - formation of solid inorganic materials from molecular precursors via room-temperature, wet-chemistry-based procedures - is easily adapted to making powders as well as films. This method typically yields amorphous TiO$_2$, and a subsequent calcination step is usually required to crystallize the material[4,5].

The anatase phase of TiO$_2$, with an optical band gap of 3.2 eV, has been proved to be the most catalytic active structure in UV light, due to its energy band structure and high surface area. On the other hand, due to their large band gap, undoped anatase, leads to low efficiency yields in solar applications. The presence of appropriate dopants revealed the possibility of using anatase structure for visible photocatalytic applications. However, it should be noticed that if the band gap is lowered beyond a limit, then the energy requirements of the photocatalytic reaction will not be achieved due to the too fast recombination electron-hole time.

The photocatalytic activity of TiO_2 in visible range can be enhanced by doping with transitional metals, which can reduce its optical band gap[6-15]. Several studies deal with experiments in which iron doped TiO_2 films have been used for various pollutant degradations. Navio et all[16] reported the photodegradation of oxalic acid under visible irradiation by using TiO_2 doped with 5 % Fe and Teoh et all[17] showed that flame spray pyrolysis Fe-TiO_2 (Fe/Ti ratio of 0.05) photocatalyst preparation have a high activity relating to oxalic acid mineralization under visible light irradiation. Phenol degradation based on Fe-doped TiO_2 samples under visible light was reported by Nahar[9] et al. A molar ratio of 0.005 Fe content was found to be the optimum concentration in this particular case. The decomposition of methyl orange under visible light irradiation has been reported by Wang[18] with the highest photocatalytic reactivity obtained for doping TiO_2 with 1 % at. Fe(III).

The absorbance in the visible region was found to increase with the amount of iron introduced in the matrix of titania, which could be an evidence for photocatalytic reactivity of TiO_2 in visible light[8]. Thus, the doping of TiO_2 with Fe leads to a shift in the UV-VIS spectrum of TiO_2 due to the contribution of the 3d electron states of Fe(III)[8].

In present study we showed that antibacterial activity of $TiO_2(Fe^{3+}, PEG_{600})$ three-layers thin films, prepared by sol-gel method, is correlated with the structure and morphology of their surfaces and in the same time is influenced by iron and PEG content from coatings compositions.

EXPERIMENTAL

The $TiO_2(Fe^{3+}, PEG_{600})$ films were prepared by sol-gel & dipping method as previously described[19,20]. The densification of the films was obtained by a thermal treatment (TT) at 500^0C for 4h. The investigated samples are described in the Table I.

Table I. Samples composition

Sample	PEG (M)	Fe (1%)	Layer numbers
P0	0	1	3
P1	0.014	1	3
P2	0.029	1	3
P3	0.069	1	3
P4	0.110	1	3
P5	0.017	7	3
P6	0.069	7	3

The crystallinity of the samples was studied by X-ray Diffraction (XRD) method. The morphology and the roughness of the surfaces were investigated by atomic force microscopy (AFM) measurements. Easy Scan 2 model from Nanosurf® AG Switzerland was employed with a high resolution scanner of 10 μm x 10 μm x 2 μm, having a vertical resolution of 0.027 nm and a linear mean free error of better than 0.6%, and by using pyramidal-shape sharp tips with a radius of curvature of less than 10 nm. SPIP™ package software (Image Metrology) was used for quantitative analysis of the AFM pictures in order to obtain statistical information regarding the roughness, the dimensions of the surface aggregates (crystallites) and self-similarity in terms of Mean Fractal Dimension. The AFM images have been recorded from a large scale (10 µm) down to the sub-micrometer scale (hundreds of nm).

The optical properties (optical constants and optical band gap) of the films were investigated by Spectroscopic Ellipsometry (SE) with a null setup in the visible spectral range (0.4 µm - 0.7 µm).

The antibacterial properties of tested films were investigated towards *E. coli* strain supplied from the collection of Institute of Biology Bucharest. The experiments were carried out under natural illumination conditions with day-night cycles. In order to estimate the effect of coatings chemical composition against microbial cells, the *E. coli* was cultivated in the presence of tested samples for 72 hours at 37^{0}C in flasks containing 22.5 ml nutrient broth and 2.5 ml bacterial inoculum, which correspond to an optical density (O.D.) of about 0.16 at the wavelength of 660 nm. The number of microbial cells in these conditions appears to be close to sample taken from some waste water treatment plants. The inhibitory effect of coatings composition on the growth of *E. coli* was evaluated by reading the optical density at 660 nm with an UV-VIS Spectrophotometer (UNICAM HELIOS α) having as reference nutrient broth. After this experiment, the samples were rinsed with sterile distilled water and transferred in a flask containing 25 ml of nutrient broth. The aim of this second experiment was to check if bacterial cells are able to show adherence to tested films due to porosity of coatings as consequence of the PEG content in films compositions.

RESULTS AND DISCUSSIONS

A previous study[19] revealed that a better photocatalytic activity of TiO$_2$ films is obtained if coating is deposited on a SiO$_2$ buffer layer and not directly on the glass substrate. Thus, the diffusion of sodium ions from substrate into the film, which could stimulate the transformation from anatase to rutile phase, is avoided. It is important not only to have a good photocatalytic activity but also to maintain it in time, thus means that the bacteria must have a good adherence on the substrate, which could be obtained when the samples are co-doped with PEG (which leads to the enhancement of the porosity).

Antibacterial tests

The influence of the surface chemical composition (from the seven samples described here) on *the growth inhibition of the E.coli* was estimated from the behavior of the temporal development curves – see Figure 1a. It should be mentioned that in contrast with our previous study[19] in this work we have used a much higher concentration of bacterial inoculum in order to simulate the antibacterial activity of the TiO$_2$(Fe^{3+}) films in condition closer to those found in wasted waters.

The results from Figure 1a showed that tested microorganisms are inhibited towards grow in the presence of investigated films, having specific evolutionary dynamic during to 72 hours of observation.

After the first experiments, the samples have been transferred in 25 ml of sterile nutrient broth, in order to investigate *the adherence of the cells of the E. coli* (figure 1b) on the tested films and to clarify the influence of the composition and films morphology on the antibacterial activity, in the presence of low number of bacterial cells. (In fact only the bacteria which adhered on the surface of the films are in this case presented in culture medium).

The influence of PEG content on bacterial growth inhibition
- Tested cells adhere to the surface of all the nano-films prepared with PEG
- A reduced intensity of the inhibitory effect is noticed for the sample prepared without PEG (P0) because the bacteria do not adhere to the surface of the films
- At the same Fe concentration the inhibitory effect generally lowers with PEG content, exception being the sample P3, which have also the best adherence

Sample P4. containing the highest PEG concentration. exhibits the lowest inhibition rate

The influence of Fe concentration on bacterial growth inhibition

By increasing the Fe content from 1 to 7% an increase of the inhibitory effect at the same PEG value (P1 and P5) was observed. By increasing the PEG content (from P5 to P6) the decrease of the inhibitory effect is noticed as well.

In fact the contributions of Fe and PEG cannot be explicitly separated. but understood in terms of optimal concentrations improving the photocatalytic act. On the other hand these data showed that porosity of films. as consequence of the presence of PEG in coating composition has an important role for bacterial adherence.

These antibacterial features are close related to morphological, structural, chemical and optical properties of the samples as a matter of course to the technological parameters of our samples. such as: number of layers. type of substrate, Fe and PEG contents.

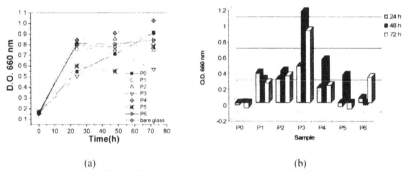

(a)   (b)

Figure 1. Growth of E. coli in the presence of coating films with different composition (a): the influence of coatings composition on the adherence of *E. Coli* to the films surface (b)

XRD analysis

From our previous study[20]. we found that in the case of deposition on glass substrate. one layer film is amorphous and two layer films have already anatase structure (crystallized in (101) direction). but three layer films have a mixed structure: anatase and rutile. To avoid the formation of rutile phase a SiO$_2$ buffer layer was introduced between glass substrate and the TiO$_2$ film. XRD have evidenced on all films studied in the present paper only the anatase phase. The crystallinity of anatase improving with the increasing of layers numbers. This improving is more pronounced when small PEG content was added (Figure 2-sample P1). In this case the anatase phase crystallizes also in (200) direction.

It should be mentioned that the anatase phase preserves more active sites in photocatalytic reactions and exhibits a remarkable reactivity in this respect as compared to rutile phase of TiO$_2$. The photocatalytic activity depends on the crystallinity of TiO$_2$. so that the increased crystallinity enhances the photocatalytic activity. However. the increase in the crystallite size decreases the specific surface area and the number of active surface sites thus lowering the photocatalytic activity. Hence. the design of nanostructured materials with small crystallite size. high surface area and open porosity is essential for the development of photocatalyticaly active materials.

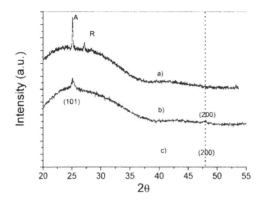

Figure 2. XRD diffractograms of three-layer films: a) P0 without SiO_2 buffer; b) P0; c) P1.
(A means anatase and R means rutile)

AFM analysis

Among the favorable features of the anatase phase which were found to play an important role in the antibacterial activity, the nanometric size of the film particles is one of the most important. In these sense, a morphological study was done by AFM in an attempt to find qualitative and quantitative details on the nanometric scale on the surface features of the $TiO_2(Fe^{3+}, PEG_{600})$ films.

Table II shows the results obtained for the roughness (expressed by Root Mean Square - RMS - values), mean and minimum crystallites diameters, averaged mean size and fractal mean dimension (MFD).

Table II. Roughness (RMS), grain diameter, mean size, mean fractal dimension (MFD) determined by AFM analysis at 1 μm scale

Sample	RMS (nm)	Grain diameter (nm)		Mean Size (nm)	MFD
		mean	minimum		
P0	0.64	19.233	10.839	24.467	2.80
P1	0.50	19.389	11.707	23.066	2.93
P2	0.66	32.744	12.516	47.178	2.96
P3	0.48	32.375	13.275	42.609	2.89
P4	0.74	20.000	11.707	25.294	2.79
P5	0.20	31.462	12.516	48.693	2.78
P6	0.70	70.523	30.778	94.951	2.87

Since the porosity of the films can be controlled by the PEG content, (which leads to a morphology which appear to support the antibacterial activity) we concluded that is important to compare the morphology of the films with similar PEG concentration but with different concentration of Fe and, accordingly, with the same Fe concentration but prepared with a different PEG amount, or, in other words, with a different morphology. Thus, Figure 3 shows in comparison 2-D (topography or contrast phase) and 3-D (topography) AFM images, at a scale of 1 μm, for the samples: P0 (a and b)– prepared without PEG; P1 (c and d) and P5 (e and f) with near PEG concentration but doped with 1% Fe^{3+} and, respectively 7% Fe^{3+}, P2 (0.029 PEG, 1% Fe) – (g and h) and P6 (0.069 PEG, 7%, Fe) – (i and j).

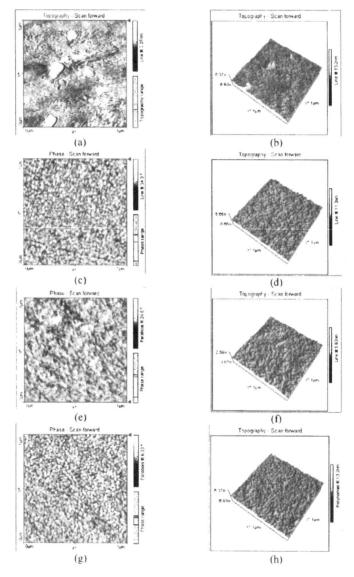

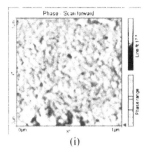
(i)

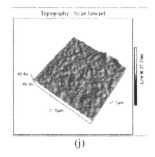
(j)

Figure 3. AFM 2-D images which exhibit the fine grain structure of the films in comparison with 3-D AFM images revealing the topography of the nanostructured, iron-doped TiO_2 films: P0 (a and b) without PEG; P1 (c and d) and P5 (e and f) – similar PEG but different Fe content; P2 (g and h) and P6 (i and j) – different PEG and Fe content.

Even if the PEG is large molecule, which after annealing leads to an increased porosity, as will be shown in details in the ellipsometry paragraph, the surface of the film prepared without PEG (P0 – a and b) appear to be defects-rich, in comparison with the film-series doped with PEG (c – j).

The mean size of the surface grains was found to be in the range between 23 nm (P1) and 95 nm (P6). It could be however remarked that the content of the Iron seems to play an important role in the surface morphology of the $TiO_2(Fe^{3+}, PEG_{600})$ films. Thus, the sample without PEG (P0) and the sample with the lowest PEG content (P1) both doped with 1% Fe^{3+}, have grains with a mean size around 23-24 nm (Figure 4). By increasing the PEG content (samples P2 and P3) the grains mean size increased twice, at least to a value of 47 nm. In turns, the maximum PEG content (sample P4) in the film-series with 1% Fe, leads to a surface morphology characterized by grains average size of 25 nm. Further, despite the very similar aspect of the surface features, as can be observed from fig. 3, the quantitative analysis of the AFM images shows a major increasing (Table II) of the surface grains for the films doped with 7% Fe. In this sense, by comparing the samples P1 and P5, having almost the same PEG content but with a distinct Fe concentration, it can be observed that the grains mean size is doubled by using a 7% iron content. Further, by doubling the PEG content (sample P6) in the preparation stage, it can be noticed the same linear tendency regarding the average size of the surface grains with a value of 95 nm. This is qualitatively visible by close inspections of the 2-D AFM images from Figure 3. It is noteworthy to mention relatively constant values for the grains minimum diameter, of the samples from Table II, which is found in the same range of 10-13 nm, except for sample P6, with a value of 30 nm. Despite of the different PEG and Iron content, which leads to a different surface morphology, the roughness of these surfaces does not present a certain behavior and have closely values around 0.5-0.7 nm, except for the sample P5, characterized by the lowest roughness (0.2 nm).

A fractal behavior on a nanometric scale range for the whole series of TiO_2 films from Table II can be mentioned, with a closely mean fractal dimension of 2.8-2.9. These values are characteristic for a surface whose roughness originates in grains dimensions and their particular spatial distribution. This could be understood as a confirmation of the surface nanoparticles formations, reflecting the self-similar character of the studied surfaces. This means that the surface of the samples exhibits self-similarity on a rather large domain (with long-range fractal behavior characterizing correlations between particles).

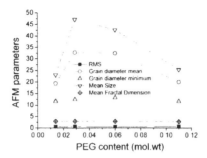

Figure 4. Variation of the AFM parameters with PEG content for a TiO$_2$ (1%Fe)

Ellipsometric characterization

Another important parameter in the anatase phase stabilization is the porosity (P) of the films which was calculated from the refractive index (n) of the film from the formula[21]:

P=1-(n^2-1)/(n_0^2-1), where n_0 is the refractive index of anatase.

The refractive index was obtained from ellipsometric analysis. The computations were done by taken into account the EMA[22] model with two layers on the substrate: TiO$_2$(Fe, PEG) film (first layer) / SiO$_2$ buffer (second layer) /glass. The ellipsometric results were shown that:

- the refractive index decreases with the PEG content (the samples remaining porous even after annealing for 4 h at 500^{0}C). At the same PEG content the refractive index increases with the Fe content (Figure 5a)

- the absorption index do not depend essentially by the PEG content, but increases with the Fe content (Figure 5b)

- the optical band gap (Eg) of the films, calculated from Tauc[23] formula based on k dispersia, with 1%Fe are: Eg(direct)=2.52-2.56 eV and Eg(indirect)=1.82 eV. The optical band gap of the films with 7% Fe are: Eg(direct)=2.64 eV and Eg(indirect)=1.7 eV

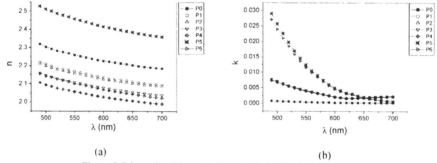

(a)   (b)

Figure 5. Dispersia of the optical constants (n, k) of the films

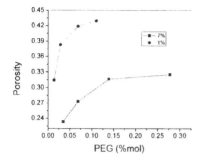

Figure 6. Porosity of the films vs PEG content

The variation of the sample porosity depending on the PEG and Iron content, respectively, is presented in Figure 6. It may be noticed that the porosity of the films increases with the PEG content and decreases with Fe amount

CONCLUSIONS

The antibacterial activity of the TiO₂ films, using a bacterial inoculum close to the concentration of microbial cells identified in wasted water was found to be enhanced by an optimal iron and PEG contents (P3 and P6). The surface films morphology for the best adherence of the cells and the most efficient antibacterial reaction is stronger at a low content of iron.

All the samples were crystallized in the anatase phase. Its stabilization is enhanced by the film porosity induced by PEG codoping as well as the nanometric size of the surface grains.

The films are very smooth, the roughness values being not higher than 0.74 nm, the grain medium diameter being in the range of 20-70 nm and the fractal dimension in 2.79-2.96 range.

The direct optical band gap values increases with the Fe content up to 2.64 eV (470 nm), making TiO₂(Fe, PEG) an active photocatalyst under visible light. The absorption index increases with the Fe content, leading to the possibility to enhance the efficiency yields in solar applications.

ACKNOWLEDGEMENTS

The financial support of CEEX-(318/2006) and CNCSIS 1276/2007 Romanian projects is gratefully acknowledged.

REFERENCES
[1]P.-C. Maness, S. Smolinski, D. M. Blake, Z. Huang, E. J. Wolfrum and W. A. Jacoby, Bactericidal Activity of Photocatalytic TiO₂ Reaction: toward an Understanding of Its Killing Mechanism, *Appl. Env. Microbiol.*, **65**, 4094–4098 (1999).
[2]D.F. Ollis, E. Pelizzetti and N. Serpone, Destruction of Water Contaminants, *Environ. Sci. Technol.*, **25**, 1523 – 1529 (1991).
[3]A.Fujishima, and K.Honda, Electrochemical photolysis of water at a semiconductor electrode nature, *Nature* (London), **238**, 37-38 (1972).
[4] A.Linsebigler,G. Lu and J. Yates, Photocatalysis on TiO2 surfaces: principle, mechanisms, and selected results, *Chemical Review*, **95**, 735-758 (1995).
[5]C.J. Brinker and G.W. Scherer, Sol-Gel Science: The Physics and Chemistry of Sol-Gel Processing, Boston: Academic Press, 1990.

[6]C. Wang, C. Bottcher, D. W. Bahnemann and J. K. Dohrmann, A Comparative Study of Nanometer sized Fe(III)-doped TiO$_2$ Photocatalysts: Synthesis, Characterization and Activity", *J. Mat. Chem.* **13**, 2322-2329 (2003).

[7]W.-C. A. Hung, S.-H. Fu, J.-J. Tseng, H. Chu and T.-H. Ko, Study on Photocatalytic Degradation of Gaseous Dichloromethane using Pure and Iron Ion-Doped TiO$_2$ Prepared by the Sol-Gel Method", *Chemosphere*, **66**, 2142-2151 (2007).

[8]C. Adan, A. Bahamonde, M. Fernandez-Garcia and A. Martinez-Arias, Structure and Activity of Nanosized Iron-Doped Anatase TiO$_2$ Catalysts for Phenol Photocatalytic Degradation, *Appl. Catal. B: Environ.*, **72**, 11-17 (2007).

[9]M. S. Nahar, K. Hasegawa and S. Kagaya, "Photocatalytic Degradation of Phenol by Visible Light-Responsive Iron-Doped TiO$_2$ and Spontaneous Sedimentation of the TiO$_2$ Particles", *Chemosphere*, **65**, 1976-1982 (2006).

[10]H. Yamashita, M. Harada, J. Misaka, M. Takeuchi, B. Neppolian and M. Anpo, Photocatalytic Dgradation of Oganic Compounds diluted in Water using Visible Light-Responsive Metal Ion-implanted TiO$_2$ Catalysts: Fe Ion-implanted TiO$_2$", *Catal. Today*, **84**, 191-196 (2003).

[11]J.O. Carneiro, V. Teixeira, A. Portinha, A. Magalhaes, P. Coutinho, C. J., Tavares and R. Newton, Iron-doped Photocatalytic TiO$_2$ sputtered Coatings on Plastics for Self-cleaning Applications, *Mater. Scn. and Eng. B: Solid-State Mater. for Advan. Techn.*, **138**, 144-150 (2007).

[12]C. Wang, Q. Li and R. Wang, Synthesis and Characterization of Mesoporous Iron-Doped TiO$_2$, *J. Mater. Scn.*, **39**, 1899-1901 (2004).

[13]D. H. Kim, H.S. Hong, S. J. Kim, J. S. Song and K. S.Lee, Photocatalytic Behaviors and Structural Characterization of Nanocrystalline Fe-doped TiO$_2$ Synthesized by Mechanical Alloying", *J. Alloy. Comp.*, **375**, 259-264 (2004).

[14]R. S. Sonawane, B. B. Kale and M. K. Dongare, Preparation and Photo-catalytic Activity of Fe-TiO$_2$ Thin Films Prepared by Sol-Gel Dip Coating, *Mater. Chem. Phys.*, **85**, 52-57 (2004).

[15]M. Kang, Synthesis of Fe/TiO$_2$ Photocatalyst with Nanometer Size by Solvothermal Method and the Effect of H$_2$O addition on Structural Stability and Photodecomposition of Methanol, *J. Mol. Catal. A: Chem.*, **197**, 173-183 (2003).

[16]J. A. Navio, G. Colon, M. I. Litter and G. N. Bianco, Synthesis, Characterization and Photocatalytic Properties of Iron-Doped Titania Semiconductors Prepared from TiO$_2$ and Iron (III) Acetylacetonate, *J. Mol. Catal.: A: Chem.*, **106**, 267-276 (1996).

[17]W. Y. Teoh, R. Amal, L. Madler and S. E. Pratsinis, Flame Sprayed Visible Light-Active Fe-TiO$_2$ for Photomineralisation of Oxalic Acid, *Catal.Today*, **120**, 203-2013 (2007).

[18]X.H. Wang, J.-G. Li, H. Kamiyama, Y. Moriyoshi and T. Ishigaki, Wavelength-Sensitive Photocatalytic Degradation of Methyl Orange in Aqueous Suspension over Iron(III)-Doped TiO$_2$ Nanopowders under UV and Visible Light Irradiation, *J. Phys. Chem.: B*, **110**, 6804-6809 (2006).

[19]C.C. Trapalis, P. Keivanidis, G. Kordas, M. Zaharescu, M. Crisan , A. Szatvanyi and M. Gartner, TiO$_2$(Fe^{3+}) Nanostructured Thin Films with Antibacterial Properties. *Thin Solid Films*, **433**, 186-190 (2003).

[20]C. Trapalis, M. Gartner, M. Modreanu , G. Kordas , M. Anastasescu , R. Scurtu and M. Zaharescu, Stabilization of the Anatase Phase in TiO$_2$(Fe^{3+}, PEG) Nanostructured Coatings, *Appl. Surf. Sci.*, **253** 367-371 (2006).

[21]B.E. Yoldas, Investigations of porous oxides as an antireflective coating for glass surfaces. *Applied Optics* **19**, 1425-1439 (1980).

[22]D.G.A. Bruggeman, Berechnung Verschiedener Physikalischer Konştanten von Heterogenen Substanzen, *Ann. Phys. (Leipzig).* **24** 636-679 (1935).

[23]J. Tauc, R. Grigorovici, A. Vancu, Optical properties and electronic structure of amorphous germanium, *Phys. Status Solidi* **15**, 627-637 (1966).

POLYETHYLENE/BORON CONTAINING COMPOSITES FOR RADIATION SHIELDING APPLICATIONS

Courtney Harrison[1], Eric Burgett[2], Nolan Hertel[2], Eric Grulke[1]
[1]Chemical & Materials Engineering, University of Kentucky, Lexington, Kentucky, USA
[2]Neely Nuclear Research Center, Georgia Institute of Technology, Atlanta, Georgia, USA

ABSTRACT
Multifunctional composites made with boron-rich additives are absorbers of low energy neutrons, and could be used as radiation shielding materials. Nanoparticle additives of boron carbide, boron nitride, and other boron-containing materials might make good multifunctional composites if they are paired with polymers having high levels of hydrogen, such as polyethylene. Polyethylene/boron containing nanocomposites were fabricated using conventional polymer processing techniques, and were evaluated for mechanical and radiation shielding properties. Addition of boron-rich nanoparticles to an injection molding grade HDPE showed superior mechanical properties to neat HDPE. Radiation shielding measurements of the nanocomposites were improved over those of the neat polyethylene with boron carbide showing the best results. Worker exposure to potentially toxic nanoparticles can be reduced by using appropriate manufacturing processes. Nanocomposites are advantageous because of their increased mechanical properties and superior shielding performance.

INTRODUCTION
The shielding of radiation for astronauts has been an important element of NASA missions for many years. NASA's current missions include a return trip to the moon by 2020 and design for a Mars expedition[1]. The atmosphere outside the Earth's orbit contains mostly protons, but other particles are present that can be dangerous to humans[2]. Galactic cosmic rays (GCRs) are a threat because 1 to 2% of these nuclei have high charge, with Z values greater than 10 and energies greater than 100GeV. Particles present in GCRs have a significant impact and have a high specific ionization capability; they produce a large number of ion pairs per unit length of their track. They can contribute to 50% of the radiation dose an astronaut would receive[3]. Solar particle events (SPEs) are also a continued threat to humans in space. Advanced shielding methods are essential to protect astronauts for extended missions.

Conventional shielding materials such as aluminum require large thicknesses to effective.Aluminum will permit an increase in cancer induction rates for relatively thick sections[4] due to secondary particles generated when neutrons and protons are stopped. Research has shown the effectiveness of the shield is directly related to the mass number: effectiveness increases as mass number decreases, with hydrogen being the best material[5]. Galactic cosmic rays can be fragmented into less damaging particles with hydrogen-containing shielding materials. Lower atomic number materials also produce lower number of secondary particles. Polymers with high hydrogen content, such as polyethylene, are advantageous because they are lightweight and could easily be fabricated into sheets for laminate, extruded into foams for insulation, or injection molded for structural parts. NASA has focused on studying polyethylene because of its high hydrogen content and has chosen it as a reference material for multifunctional composites currently being developed[6].

Multifunctional polymer composites have many advantages for radiation shielding. Inorganic fillers can improve mechanical strength, radiation shielding and even flame retardancy properties. These fillers also need to have low atomic mass numbers to stop incoming radiation and to produce fewer secondary particles. Boron-containing inorganic fillers are attractive for these reasons. In particular, boron carbide (B_4C) and boron nitride (BN) could be used as fillers into a polyethylene matrix to create multifunctional composites. Studies of these materials have recently been completed to determine their effectiveness with micron sized particles[7,8]. This paper will focus on preparing multifunctional nanocomposites for radiation shielding applications.

Nanocomposites are advantageous because they can provide significant improvements in mechanical properties. Recent studies have shown that an inorganic nanophase/polymer system can demonstrate improved mechanical properties. Micron sized additives can reduce the mechanical strength when poor adhesion exists between the particles and continuous phase while nanoparticles can give improved properties at low loadings of 1-5wt%. Nanoparticles such as clay[9], calcium carbonate[10], and hydroxyapatite[11] have been added to high density polyethylene and shown increased mechanical properties.

The purpose of this work is to develop multifunctional nanocomposites for radiation shielding applications. The filler will be inorganic and also nanoparticles of boron nitride and boron carbide. The continuous phase used will be high density polyethylene because of its high hydrogen content.

EXPERIMENTAL MATERIALS AND METHODS

High density polyethylene was purchased from Dow Chemical Company (Houston, TX), grade NT 8007, an injection molding grade. Boron carbide, supplied by Nanostructured and Amorphous Materials, Inc. (Houston, TX), was used as received with an average particle size of 50 nm. Boron nitride was also purchased from Nanostructured and Amorphous Materials, Inc. with an average particle size of 137 nm and was used as received. Particle sizes were determined by TEM measurements.

Composite Preparation and Mechanical Testing

Composites were melt blended as reported previously with micron size particles of boron nitride and boron carbide[7,8]. Polymer was melted in a Haake Rheomix (Waltham, MA) melt blender at 145°C for 30 minutes. The powder samples were slowly added to the molten material with mixing. The materials were removed, allowed to cool and then crushed using a hammermill crusher into sizes of less than 0.5 mm. The crushed material pressed into thin films at 150 °C for 15 minutes using a Carver Lab Press Model C (Wabash, IN). The films were 0.002" thick and dog bone samples were cut using an ASTM D638 Type IV Die. The mechanical properties were determined on these samples using a MTS QTest 10 (Eden Prairie, MN) with a strain rate of 0.0167 sec^{-1} and repeated five times. The standard error for tensile modulus was 130.4 MPa. The strains at break values have a standard error of 7.9% and tensile strength has an error of 4.5 MPa.

Neutron and Proton Attenuation Measurements

The materials were pressed into solid plaques approximately 11.25 x 7.95 x 0.5 cm thick and tested against aluminum (1100 grade), a conventional material used for space radiation shielding. Composites were tested using previously developed methods[7,8] at both the Los Alamos Neutron Science Center (LANSCE) Weapons Neutron Research (WNR) neutron facility as well as the Fermi National Accelerator Lab (FNAL). At LANSCE, the samples were tested in the WNR beam on the 30 left flight paths (FP30L). The neutron beam at FP30L was selected for neutron testing because it delivers up to 600 MeV neutrons. A similar energy spectra is shown in Ferenci and Hertel's work[12]. The neutrons are produced by spallation events caused by an 800 MeV pulsed proton beam incident on a tungsten target. These spallation neutrons have a very similar energy spectrum to that of neutrons at 40,000 feet, an area of interest for high altitude flight and low earth orbit. At FNAL, the M02 beam line was used to deliver 120 GeV protons to measure attenuation through the samples. 120 GeV protons were selected because of their ease of extraction from the Main Injector, and because their energy level is near the maximum energy flux of galactic cosmic radiation (GCR).

A water tank phantom was used to measure the shielding effectiveness and the setup is shown in Figure 1. Absorbed dose measurements were made at various depths with several different thicknesses of shielding material to determine the relative shielding effectiveness. If polyethylene-based composites can perform as well or better than the aluminum, then the polyethylene composites are a superior choice. Polyethylene compounds, comprised of only light nuclei, have a much lower neutron yield in high energy particle reactions. A tissue-equivalent ion chamber made out of A150 tissue-equivalent plastic was used to measure the absorbed dose in the tank. This integral measurement of dose allows a direct measure of shielding effectiveness. Due to the fluctuations of these beams, beam monitors were used to normalize the results.

Figure 1. Setup of water tank phantom at Fermilab labeled

RESULTS AND DISCUSSION

Mechanical Properties

Tensile tests were performed on the nanocomposites and tensile strength, modulus and strain at break were recorded. Boron nitride and boron carbide nanoparticle composites were compared to that of the HDPE. An averaged stress versus strain curve is shown for five replicates of three different volume percent of boron carbide in Figure 2. The HDPE has an elongation at break value of approximately 300% and this plot shows a value of only 14% for the 1 vol% boron carbide nanocomposite. The addition of the filler reduces the elongation at break values because the particles are acting as defect sites. This defect causes failure to occur in the polymer matrix. The plot shows with increasing amount of filler present in the composite the quicker the sample fails.

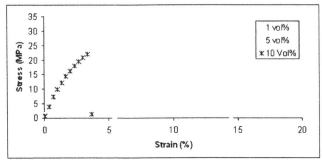

Figure 2. Averaged stress versus strain curve for 1, 5, and 10 vol% boron carbide filler

Elongation at break values for the boron nitride nanocomposites show a similar trend to those seen in boron carbide. A plot of those values is seen in Figure 3. The elongation at break value for 1 vol% boron nitride nanocomposite has a value of 25% which is larger than the boron carbide nanocomposite. However, defects created by the addition of particles still greatly reduces the value from that seen with HDPE.

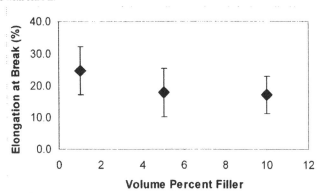

Figure 3. Elongation at break values for increasing volume percent of boron nitride filler

Tensile modulus values are shown in Figure 4 and an increase is seen for both of the nanocomposites over that of the neat polyethylene. At low volume percents, both of the boron fillers have similar values. As the amount of filler increases, the boron carbide composites show a faster increase than the boron nitride composites. Tensile modulus is affected by the amount of filler present. It is thought that at higher loading, when more particles are present, particle-particle interactions dominate the fracture mechanics causing the increase [13].

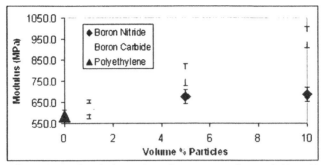

Figure 4. Tensile modulus values for increasing volume percent of boron nitride and boron carbide

Tensile strength values are shown in Table 1 for both composites. The neat HDPE has a tensile strength of 29.5 MPa and at the lowest loading levels, both of the composites show an increase with boron nitride showing a higher value than boron carbide. However, with an increase in filler in the composite the values are decreased for the B_4C while the BN remains unchanged. The decrease can be explained for boron carbide by the particle debonding from the matrix at higher loading levels. With more particles present, the particles created more voids, which lowered the tensile strength. Atikler saw similar results with the addition of fly ash/HDPE composites [13]. For boron nitride, however, the decrease was not seen. The particles had strong enough adhesion to the matrix that they did not debond and cause the decrease.

Table 1. Tensile strength values for boron nitride and boron carbide composites, MPa.

Volume Percent	Boron Nitride	Boron Carbide
0	29.5	29.5
1	33.1	30.9
5	33.3	24.8
10	33.4	22.8

Radiation Shielding Properties

Shielding effectiveness of the composites was measured by the dose transmission. This is defined as the ratio of the dose rate with shielding material to the dose rate with no shielding material. The larger the slope in the plot of dose transmission versus thickness of shielding material indicates the best material at attenuating the radiation. Results for the boron nitride nanocomposites from the WNR neutron beam are shown in Figure 5. The transmission factors are similar for aluminum, polyethylene, and the composites. Under high energy neutron radiation, polyethylene-based materials have the advantage of not producing long lived radioactive progeny that could produce more radiation. The aluminum counterpart produced longer lived progeny that yielded high energy beta and photon radiation as a result of neutron capture.

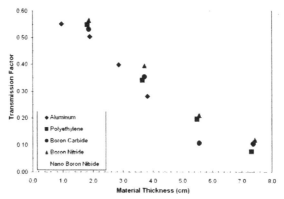

Figure 5. WNR neutron beam showing transmission factor versus material thickness for Al. HDPE. and PE/boron composites

Transmission factors from the Fermilab 120 GeV proton beam for the various samples are shown in Figure 6. The nanocomposites with increasing thickness showed slightly better shielding effectiveness than the HDPE. After 1.5 cm of shielding material. both show little change. This is due to spallation that could occur at such high energies. A higher dose was recorded because the polyethylene based materials actually caused more spallation products to enter the tank. For high energy protons. aluminum appears to be the best material. It should be noted that at 120 GeV. it takes several meters of solid steel to bring these energetic particles to rest.

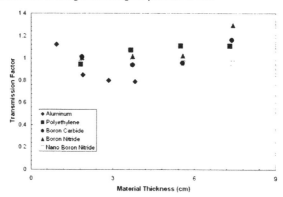

Figure 6. 120 GeV proton beam showing transmission factor versus material thickness for Al. HDPE, and PE/boron composites.

Boron carbide nanocomposites were not tested for their shielding effectiveness at these two sites. The data for the boron nitride composites for both nanoparticles and micron sized particles was included in Figure 5 and Figure 6. From these two experiments. it is clear that the nanoparticles and micron size particles behave very similarly for shielding effectiveness. The only noticeable difference in the data is with transmission factor of the high energy proton beam shown in Figure 6. the nanocomposites appear to have increased shielding capabilities with increasing thicknesses. This could be due to the smaller particles having better dispersion and providing a better shield than their larger sized counterparts. The micron sized data for boron carbide is included and it is expected that the nanocomposite will behave similarly. The boron carbide composites show similar results as that of the boron nitride in shielding effectiveness.

CONCLUSION

Advanced shielding materials are necessary for the radiation and performance challenges that will be faced with a return trip to the Moon or an initial venture to Mars. Conventional materials such as aluminum tend to be too heavy and provide large amounts of secondary radiation. Using a multifunctional composite is advantageous because it is lightweight, and provides radiation protection while reducing the formation of secondary particles.

Polyethylene based composites are the best choice because of their high hydrogen content. Boron filler improves its shielding characteristics. These composites show an increase in mechanical properties over that of the neat polymer. Mechanical properties show increases in modulus values for both composites. The boron carbide composites have better moduli due to more particles being present and interacting in the composite. Tensile strength dropped with increased particle loadings, due to debonding from the matrix. The boron nitride composites showed lower modulus values indicating less particle-particle interactions. These composites had better adhesion between the particles and the matrix, the particles did not debond, and the tensile strength was relatively unchanged from the neat polymer.

Radiation shielding effectiveness is similar for the boron composites compared to that of the neat polymer. In the neutron beam the polyethylene based composites showed an advantage with the boron nitride nanocomposite showing an improvement. In a neutron beam these composites are also expected to not produce as much secondary radiation as a conventional shielding material such as aluminum. The high energy proton beam shows that aluminum is the better shielding material and spallation was seen in the polyethylene based materials that increased the dose measured.

Multifunctional composites are beneficial because of their flexibility. They have shown to have increased mechanical properties with similar shielding effectiveness to materials currently being used.

REFERENCES

[1]   NASA, February, 2004.
[2]   J. A. Simpson, *Ann. Rev. Nuc. Part. Sci.*, **33**, 323 (1983).
[3]   R. Kiefer, D. McGlothlin, J. Chapman, and S. Thibeault, *Abst. Pap. Am. Chem. Soc.*, **225**, U558 (2003).
[4]   J.W. Wilson, F. A. Cucinotta, J. Miller, J.L. Shinn, S.A. Thibeault, R.C. Singleterry, L.C. Simonsen, and M.H. Kim, presented at the Materials Research Society Symposium Proceedings, 1999 (unpublished).
[5]   C. Zeitlin, S.B. Guetersloh, L.H. Heilbronn, and J. Miller, *Nuc. Instr. Meth. Phy. Res. B*, **252**, 308 (2006).
[6]   S. Guetersloh, C. Zeitlin, L. Heilbronn, J. Miller, T. Komiyama, A. Fukumura, Y. Iwata, T. Murakami, and M. Bhattacharya, *Nuc. Inst. Meth. Phy. Res. B*, **252**, 319 (2006).
[7]   C. Harrison, S. Weaver, C. Bertelson, E. Burgett, N. Hertel, and E. Grulke, *J. App. Poly. Sci.*, **In Publication,** (2008).
[8]   C. Harrison, E. Burgett, N. Hertel, and E. Grulke, presented at the STAIF 2008, Albuquerque, NM, 2008 (unpublished).
[9]   M. Tanniru, Q. Yuan, and R. Misra, *Poly.*, **47**, 2133 (2006).

[10]   A. Lazzeri, S. Zebarjad, M. Pracella, K. Cavalier, and R. Rosa, *Poly.*, **46,** 827 (2005).

[11]   R. Shahbazi, J. Javadpour, and A. Khavandi, *Adv. Appl. Cer.*, **105,** 253 (2006).

[12]   M. Ferenci and N. Hertel, *Rad. Pro. Dosim.*, **107,** 213 (2003).

[13]   U. Atikler, D. Basalp, and F. Tihminlioglu, *J. Appl. Poly. Sci.*, **102,** 4460 (2006).

SYNTHESIS AND OPTICAL PROPERTIES OF SiC$_{nc}$/SiO$_2$ NANOCOMPOSITE THIN FILMS

A. Karakuscu[1], R. Guider[2], L. Pavesi[2] and G. D. Sorarù[1]

aylin.karakuscu@ing.unitn.it, guider@science.unitn.it, pavesi@science.unitn.it, soraru@ing.unitn.it
[1]Department of Materials Engineering and Industrial Technology.
Engineering Faculty-University of Trento, Via Mesiano 77, 38050 Trento, Italy
[2] Department of Physics. Science Faculty-University of Trento
Via Sommarive 14, 38050 Povo, Trento, Italy

ABSTRACT

SiOC thin films are produced by the polymer pyrolysis method from sol-gel derived precursors. The starting solution is a mixture of triethoxysilane (T^{11}) and methyldiethoxysilane (D^{11}) with a T^{11}/D^{11} molar ratio of 2. Thin films are deposited on SiO$_2$ substrate by spin coating and pyrolysed in a carbon furnace under Ar flow at temperatures in the range 700–1250°C. The surface properties of SiOC films are investigated by SEM, AFM and XPS. The produced films are crack free and homogeneous even at high pyrolysis temperatures. Profilometer measurements were performed to investigate the film thicknesses. The nature of the chemical bonds present in the film structure and its photoluminescence properties are studied by Attenuated Total Reflection Fourier Transform Infrared Spectroscopy (ATR-FTIR) and Photoluminescence (PL) measurement, respectively. PL spectra change with the temperature: at low pyrolysis temperature a blue peak at around 425 nm due to the presence of defects in the amorphous structure is observed. At high temperature (1200°C) an intense yellow-green emission at 560 nm (2.25 eV) dominates the spectrum. This peak emission is assigned to SiC$_{nc}$ formed "in situ" from the SiOC phase. The results of this study show that SiOC nanocomposite thin films with *in situ* grown SiC$_{nc}$ have very promising photoluminescence properties.

INTRODUCTION

SiC$_{nc}$/SiO$_2$ nanocomposites are of great interest in the production of Light-Emitting Diode (LED) due to the emission of SiC nanocrystals in the visible spectral range from blue to yellow. Optically, bulk SiC shows a weak emission at room temperature due to its indirect band gap [1]. However, the emission intensity can be significantly enhanced when the crystallite size decreases to few nanometers [2, 3]. Accordingly, studies on the synthesis of blue-emitting SiC nanocrystals by C ion implantation have been reported [4, 5, 6]. However, these processing methods require complicated equipments and moreover the observed luminescent properties of SiC nanocrystals are not very reproducible and strongly depend on the fabrication methods. In addition, the fabrication and luminescent properties of nanocrystalline SiC films are even less studied [7]. On the other hand, polymer pyrolysis for the deposition of amorphous SiOC thin films is a simple processing route to produce nanostructured multicomponent ceramics with high thermal stability [8, 9]. According to this method, pre-ceramic thin films are converted to ceramic coatings by a pyrolysis stage in controlled atmosphere at 800-1000°C. At higher temperatures the partitioning of SiOC phase leads to the *"in situ"* crystallization of nano-scale SiC into an amorphous SiO$_2$ matrix. The nanostructural features such as nanocrystal size and amount can be easily controlled through a number of processing parameters such as the composition of the preceramic film, the heat treatment temperature, atmosphere and time [10].

In this study, SiC$_{nc}$/SiO$_2$ nanocomposite thin films were synthesized from sol-gel-derived precursors by pyrolysis in Ar flow at different temperatures up to 1250 °C. Structural change and optical properties of the SiOC films were investigated as a function of the pyrolysis temperature.

EXPERIMENTAL

Gel films were deposited on silica substrates by spin coating. The starting solution was prepared by mixing triethoxysilane (T^H) HSi(OEt)$_3$, and methyldiethoxysilane (D^H), HCH$_3$(OEt)$_2$ using ethanol as solvent. A mole ratio of $T^H/D^H=2$ was chosen to allow the formation of a stoichiometric SiOC glass without excess carbon [11]. The alkoxides solution was pre-hydrolized with acidic water (pH =4.5, HCl) using a H$_2$O/OEt ratio of 0.5 to allow enough time for film production and to promote a slow gelation. The solution was spun at 4000 rpm for 1 minute on ultrasonically cleaned SiO$_2$ (Heraeus -HSQ300) substrates. The gel films were stabilized at 80°C for 24 hours before pyrolysis. The pyrolysis process was carried out in a C-furnace under Ar flow (100 ml/min) with a heating rate of 5°C/min at different temperatures, in the range 700-1250 °C with 1 hour holding time at the maximum temperature.

Film thicknesses during the pyrolysis treatment were measured by using a Hommel tester T8000 profilometer. A scratch was introduced in the gel film and the depth of the scratch measured by the profilometer. Several scans ware made across the step and the film thickness was determined as an average of about 6 measurements over a length of 1.5 cm. The error coming from instrument is taken as ± 5 %.

Surface quality of the films was investigated by Jeol JSM-5500 Scanning Electron Microscopy (SEM) and NT-MDT P47H Atomic Force Microscopy (AFM), with a 50X50 μm scanner range. Chemical composition of the films was examined by X-ray photoelectron spectroscopy (XPS), using a SPECS analyser Phoibos100 with 5 channeltrons detection working at 2.10-10 Torr, equipped with Mg Kα source (1253.6 eV) Xray gun which is non monochromated. In order to follow the structural evolution occurring during the pyrolysis process, a Nicolet Avatar 330 instrument was used in reflectance mode to collect Fourier-Transform Infrared-Attenuated Total Reflection (FTIR-ATR) spectra. The FTIR spectra were collected from 4000 to 600 cm^{-1} and, to improve the signal quality, 256 scans were performed for each analysis. Optical properties of the films were studied by photoluminescence measurements (PL) and spectra were recorded at room temperature using a spectrometer operating with an Argon Laser (365nm) as excitation source. The fluorescence light was collected from the front face of the samples in reflection mode. The power of the excitation source on the sample was 2 mW. Interferential filters were used to select the excitation wavelength in order to minimize the scattered stray light. All the spectra were corrected for the wavelength dependent response of the detection system.

RESULTS AND DISCUSSION

The evolution of the thickness and the shrinkage of the film with the pyrolysis temperature are reported in Figure 1. The thickness of the "as deposited" coating is 1.2 μm and decreases to 420 nm at 1250 °C. The thermal evolution weight loss and shrinkage of bulk pre-ceramic gels of similar composition has been already studied and it showed that the conversion process ends around 1100°C [11]. Accordingly, the thickness of the film continuously decreases up to 1100°C and then remains constant for higher pyrolysis temperatures. The maximum linear shrinkage is 65%, close to the typical values reported in the literature for polycarbosilane-derived SiOC thin films [10].

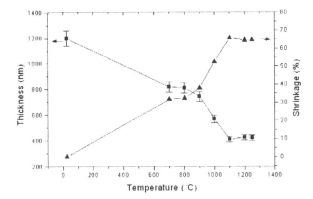

Figure 1. Evolution of the thickness and corresponding shrinkage of the film with the pyrolysis temperature.

Surface morphology and of the films were examined by SEM and AFM. SEM analysis shows that at every pyrolysis temperature homogeneous and crack free films are produced even at the edges. Figure 2 presents AFM images of the surface morphology of the film pyrolysed at 1200 °C. Average roughness is measured from the AFM images and falls in the range 0.5 – 2.5 nm. The roughness shows with the temperature a similar evolution as the shrinkage: it increases up to 1100°C and then it stabilizes.

Figure 2. AFM image of the thin film pyrolysed at 1200 °C.

In order to obtain more information about the surface properties and composition, the XPS measurements are performed and elemental compositions of the films pyrolysed at different temperatures are given in Table 1. The chemical composition of the films is close to the one measured by conventional chemical analysis of bulk SiOC samples obtained from the similar sol-gel precursor [12].

Table 1. XPS elemental composition of the investigated Si-O-C glasses

Temperature (°C)	Si (at. %)	C (at. %)	O (at.%)	Empirical formula
1000	56.4	10.9	32.7	$SiC_{0.33}O_{1.72}$
1100	57.5	10.7	31.8	$SiC_{0.34}O_{1.81}$
1200	57.6	10.1	32.3	$SiC_{0.31}O_{1.78}$

The structural evolution of the SiOC films from as-deposited to pyrolysed stage was followed by FTIR measurements. Accordingly, FTIR-ATR spectra of the as-coated and pyrolysed samples at different temperatures were collected and shown in Figure 4.

In the spectra of as-coated film (Figure 4-a), Si-H bonds of the T^H precursor gave rise to peaks at 2243 cm^{-1} and 831 cm^{-1} while Si-H bonds of D^{II} units led to a corresponding band at 2173 cm^{-1}[13]. Si-CH$_3$ stretching and rocking vibrations were revealed by the peaks at 1261 cm^{-1} and 760cm^{-1}. Peak at 1065 cm^{-1} with a shoulder at 1140 cm^{-1} was assigned as Si-O bonds in the siloxane network.

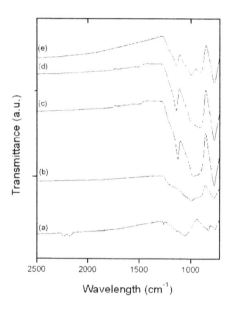

Figure 3. FTIR-ATR spectra of (a) as-coated, and pyrolyzed SiOC thin films at: (b) 800-900 °C, (c) 1000-1100 °C, (d) 1200 °C and (e) 1250 °C.

At 800-900°C the Si-H and Si-CH$_3$ related bands at 2200-2150 and 760 cm^{-1} respectively are absent suggesting that the polymer-to-ceramic transformation is complete. At the same time a complex and broad absorption band is observed in the range 1300-900 cm^{-1} with a local peak at 1000 cm^{-1} assigned to the vibration of Si-O bonds. It is assumed that this broad band includes contribution of Si-O and Si-C bonds. The presence of inorganic Si-C bonds similar to those of silicon carbide, is also suggested by the formation of a new peak at 780 cm^{-1}.

At higher pyrolysis temperatures (spectrum c) the broad peak around the Si-O vibration becomes more defined suggesting that an organization process of the SiOC network is active in this temperature range. In particular a peak appears at 1140 cm^{-1} which is assigned to Si-O vibration, the main absorption around 1000 cm^{-1} reveals a shoulder at 920 cm^{-1} related to Si-C vibrations and finally the Si-C absorption at 780 cm^{-1} becomes more intense suggesting the formation of new Si-C bonds. Above 1100°C (spectra d and e) the FTIR investigations do not show any major evolution: only the shoulder at 920 cm^{-1} grows as an individual peak, indicating a further increase the number of Si-C bonds or ordering of amorphous network [14].

PL spectra of the films pyrolysed at different temperatures are given in Figure 4. The behaviour of the films at low temperatures (800 °C-1000 °C) is different from those at high temperatures (1100 °C-1250 °C), so their spectra are given separately (4(a) and 4(b), respectively). At 700 °C, no apparent band is observed. Starting from 800 °C, films show a strong blue PL peak at around 425 nm with a broad blue-green component from 450 to 650 nm. As the pyrolysis temperature increases to 900 and then to 1000°C, the intensity of the peak at 425 nm is greatly reduced and the intensity of the second yellow PL peak is increased with a maximum at around **560** nm. In the temperature range 800-1000°C the PL band at 425 nm did not show any shift but just its intensity decreased. This observation supports the idea that it originates from defect-related centers, since it is well known that high temperature annealing can quench defects. Indeed, SiC and SiOC amorphous ceramics obtained by polymer pyrolysis at temperatures in the range 800-900°C contains a high concentration of structural defects such as free radicals. The increase of the pyrolysis temperature leads to a stabilization of the amorphous ceramic structure and to a decrease of the free radical centers [15]. This structural evolution is in agreement also with the FTIR spectra reported in Figure 3 which showed an ordering of the structure above 900 °C.

The further increase of the pyrolysis temperature above 1000 °C leads to an increase of the yellow peak intensity at 560 nm (2.25 eV) (Figure 4-b). This PL peak can be attributed to the precipitation, from the SiOC network, of C-rich clusters which could lead to the nucleation, at T≥1200°C of 3C-SiC nanocrystals. Indeed, the presence of SiC nanocrystals in bulk SiOC samples pyrolyzed at T≥1200°C and obtained from the same sol-gel solution used for the film synthesis is detected by XRD (not shown here). The XRD spectrum, in agreement with previous data [10], shows a broad peak at 2θ ≈ 35° assigned to the (111) reflection of 3C-SiC nanocrystals with an average size of 1.5 nm. Additionally, bulk samples show the same PL properties (shape and band wavelength) as films but their PL peak intensities are found significantly lower. Accordingly, we can propose that the yellow PL band at 560 nm, which starts to be visible for our SiOC films pyrolyzed at 1000°C, is due to the incipient crystallization of SiC$_{nc}$ and the formation of the expected SiC$_{nc}$/SiO$_2$ composite film. At 1250°C an abrupt decrease of the intensity of PL spectra has been observed. The reason for this result is still under investigation.

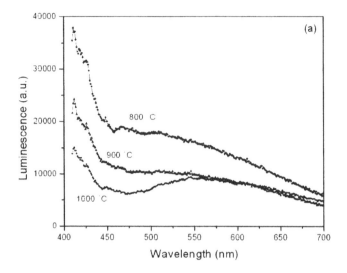

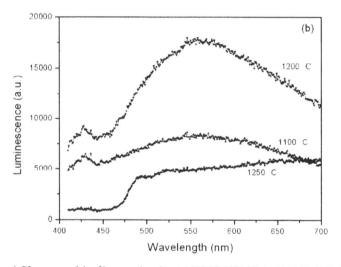

Figure 4. PL spectra of the films pyrolysed at (a) 800 °C-1000 °C (b) 1100 °C-1250 °C

CONCLUSION

SiC$_{nc}$/SiO$_2$ thin films on silica substrate were prepared by pyrolysis in inert atmosphere of sol-gel-derived coatings. Films with a thickness of 400 nm were obtained and their composition was studied by XPS technique. The chemical composition of SiOC films is close to the composition measured for bulk SiOC glasses obtained from the similar sol-gel precursors. FTIR study indicated the formation of a disordered SiOC structure at 800-900°C and an ordering process above these temperatures. At 1200 °C FTIR spectra indicate the presence of Si-O and Si-C bonds. At low pyrolysis temperature (800-900 °C) PL spectra showed a blue peak at 426 nm whose intensity decreases at high temperature and vanishes completely above 1000°C. This peak has been assigned to defects, such as free radicals, which are known to be present in high concentration in polymer-derived ceramics at intermediate pyrolysis temperatures.

Above 1000°C and up to 1200°C a broad and intense PL yellow-green peak has been observed around 560 nm. This emission has been assigned to the formation of the C-rich clusters which lead, at T≥1200°C, to the nucleation of the expected 3C SiC nanocrystals from the SiOC matrix. The results show that SiOC nanocomposite thin films with *in situ* grown SiC$_{nc}$ have very promising photoluminescence properties.

ACKNOWLEDGEMENT

This Research is supported by European Community FP6 through MCRTN-019601, PolyCerNet. The authors would like to thank to Dr Cristina Fernandez-Martin for XPS measurements and Dr Philippe Dibandjo for his interest.

REFERENCES

1. R.P. Devaty, W.J. Choyke, *Phys Stat Sol (a)* **162** (1997), p. 5
2. L.T. Canham, *Appl Phys Lett* **57** (1990), p. 1046
3. A.G. Cullis, L.T. Canham and D.J. Calcott, *J Appl Phys* **82** (1997), p. 1
4. L.S. Liao, X.M. Bao, Z.F. Yang, N.B. Min, *Appl Phys Lett* **66** (1995) p.2382
5. J. Zhao, D.S. Mao, Z.X. Lin, B.Y. Jiang, Y.H. Yu, X.H. Liu, *Appl Phys Lett* **73** (1998) p.1838
6. X.L. Wu, G.G. Siu, M.J. Stokes, D.L. Fan, Y. Gu, X.M. Bao. *Appl Phys Lett* **77** (2000) p.1292
7. W.J. Choyke, Z.C. Feng, J.A. Powell *J Appl Phys*, **64** (1988) p.3163
8. G. D. Sorarù, S. Modena, P. Bettotti, G. Das, G. Mariotto, L. Pavesi, *Appl. Phys. Lett.*, **83** (2003), p. 749
9. G.D. Sorarù, Y. Zhang, M. Ferrari, L. Zampedri, R.R. Gonçalves, *J. Eur. Ceram. Soc.*, **25** (2005) p.277
10. P. Colombo, T.E. Paulson, C.G. Pantano. *J. Sol.-Gel. Sci. Tech.* **2** (1994) p. 601
11. G. D. Sorarù, G. D'Andrea, R. Campostrini, F. Babonneau, *Proc. of the Inter. Symp. on Sol-Gel Science and Tech., ed. Am. Ceram. Soc., Sol-Gel Science and Technology*, (1995) p.135
12. G.D. Sorarù, S. Modena, E. Guadagnino, P. Colombo, J. Egan, C. Pantano, *J. Am. Ceram. Soc.*, **85** (2002) p.1529
13. G.D. Sorarù, G. D'Andrea, R. Campostrini, F. Babonneau, *J. Mater. Chem.*, **5** (1995) p. 1363
14. J.Y. Fan, X.L. Wu, P. K. Chu, *Progress in Materials Science* **51** (2006) p. 983
15. G.D. Sorarù, F. Babonneau and J.D. Mackenzie, *J. Mat. Science*, **25** (1990) p. 3886.

STRENGTH AND RELATED PHENOMENON OF BULK NANOCRYSTALLINE CERAMICS
SYNTHESIZED VIA NON-EQUILIBRIUM SOLID STATE P/M PROCESSING

Hiroshi Kimura
Department of Mechanical Engineering, School of Systems Engineering, National Defense Academy,
Yokosuka, Kanagawa, 2398686 Japan

ABSTRACT
    This article overviews the mechanical characterization of nanocrystalline ceramics synthesized by
electric current consolidation of the mechanochemically synthesized amorphous powder in the absence
of sintering additive. Berkovich indentation equipped with a depth-sensing system enables to obtain a
local yield stress under constrained compression and the strain rate sensitivity exponent (m) as an
accepted parameter of micro-plasticity; for full-density nanocrystalline $(ZrO_2)_{80}(Al_2O_3)_{20}$, the σ_y is 4.4
GPa when non-strain hardening is assumed, and the m is 0.014; these low values suggesting the
appearance of non-homogeneous flow in a nearly ideal plastic solid. For monolithic tetragonal
$(ZrO_2\text{-}3mol\%Y_2O_3)_{80}(Al_2O_3)_{20}$ with the average crystallite size of 40 nm, the flexural strength shows a
size effect by the aspect ratio of sample width (W) to height (h) having nearly 2 GPa at the maximum in
the case of $W/H \approx 1$, and a proportionality to the fracture toughness, as obtained from the
indentation-microfracture method, with the avoidance of a trade off relationship. With isothermal
superplastic forming up to a compressibility of 0.75, the m is given by a function of strain rate reduced by
Zener-Hollomon parameter (ZH) and reciprocal stress, $\dot{\varepsilon}\,\sigma^{-1}\exp(Q/kT)$ with the apparent activation
energy (Q) of 312 kJ mol^{-1}, having 0.7 and 0.3 at its higher and lower ZH respectively for nanocrystalline
$(ZrO_2\text{-}3mol\%Y_2O_3)_{80}(Al_2O_3)_{20}$.

INTRODUCTION
    The truly strong and tough ceramics is one of the most promising materials by which one can realize
various technologies of next generation that is necessary to solve today's global problems of energy
consumption and ecological destruction. The approach to the development of structural ceramics as
advanced in a few decades consists of submicron sized particle production, high densification processing,
microstructural controls with stress induced phase transformation, nanocomposite synthesis and grain
boundary modification, and has successfully brought a greatly enhanced strength. However, whether
structural ceramics can provide the reliability on strength and a practical condition for high-strain-rate
superplastic forming is still a critical issue. In order to provide a route to the innovation via
nanocrystalline synthesis in structural ceramics[1], the author has so far obtained a full-density
nanocrystalline ceramics with the average crystallite size less than a few ten nanometers by pulse electric
current consolidation of the amorphous ceramics without sintering additive[2][3], and concomitantly
found large microplasticity at ambient temperature[1], high-strength and high toughness[4], and
high-strain rate superplastic flow at low temperatures[5]. The author here is going to make clear the nature
of the strength and related phenomena inherent to the nanocrystalline structure in ceramics synthesized
via non-equilibrium solid state P/M processing with special attention to their quantitative description.

MODEL NANOMECHANICS
    A unique mechanical property of nanocrystalline ceramics such as high strength and low temperature
high-speed superplastic flow basically comes from a quasi-two-phase structure consisting of nanoscale
perfect crystal and intercrystalline amorphous phase. Furthermore, a variety of phenomena related to
strength would appear in nanocrystalline ceramics with metastable phase and nanomultiphase as
synthesized via non-equilibrium P/M processing. There is therefore a need to set up a new approach to

treat the flow and fracture in nanocrystalline ceramics as produced in the absence of sintering additive quantitatively. For the amorphous alloy recognized as an ideal plastic solid, the author has previously proposed a model mechanics study to systematize the actual Table1. Testing methods, parameters and newly suggested formula for the strength and its related phenomena of nanocrystalline ceramics as synthesized via non-equilibrium solid state PM method.

Nano-phenomenon	Testing method	Material parameter		Newly suggested formula
		Macro.	Local	
Micro-plasticity	Nanoindentation	α, Hv, ε	σ_{y}, E, n, m	$\varepsilon = A(\sigma/DH)^{1/m}$, $m = 0.014 \ll 1$
High strength & toughness	Three-point bending	σ_{f}, E	σ_{f}	$\sigma_{F} = f(W/h)$
	IM method	K_{IC}		$K_{IC} = \beta\sigma_{F}$
High-speed superplastc flow	Compression	φ, ε, m	Q, p	$m = f[\varepsilon \exp(Q/kT)]$

IM: indentation-microfracture, α: plasticity, φ: compressibility, DH: dynamic hardness

strength and its related phenomena such as extremely non-homogeneous plastic flow and fracture [6][7], and in nanocrystalline materials, is in attempt to construct a model nanomechanics by variable and parameter with a clear physical meaning, in which one can correlate the macroscopic property to nanostructure parameter. Table 1 summarizes material testing with well-defined parameter and newly suggested formula for unique phenomena in nanocrystalline ceramics of this study.

MICROPLASTICITY AT AMBIENT TEMPERATURE

Berkovich indentation equipped with depth-sensing and controls of loading and displacement rate, as is called by nanoindentation, can be used to evaluate the microplasticity at ambient temperature in nanocrystalline ceramics. Figure 1 shows the specimen surface of fully dense tetragonal $(ZrO_2\text{-}3mol\%Y_2O_3)_{80}(Al_2O_3)_{20}$ with the average crystallite size of approximately 20 nm, as prepared by pulse electric current consolidation of the ball milled amorphous powder, together with Berkovich indent and its depth profile under an applied load of 19.8 N. In this case, the nanoindentation accompanies a pile up at the edge of indent at the corner; its amount is used to obtain a relatively small value of approximately 0.1 for a strain hardening exponent (n), indicating that nanocrystalline ceramics is a nearly non-strain hardening solid. Figure 2 shows the depth-load curve for fully dense nanocrystalline $(ZrO_2)_{80}(Al_2O_3)_{20}$ specimen under a relatively low loading rate of 14 mN s^{-1}. This indentation measures in real time (i) a smooth elastic-plastic displacement up to d_1 under loading, (ii) a time dependent displacement under a constant load($d_d=d_t\text{-}d_1$, d_t is the total displacement, and d_2 is the displacement after 60 s), (iii) a elastic displacement during unloading and (iv) a residual depth (d_p). The plasticity (α) is commonly defined by $100d_p/d_t$, and gives a relatively high value of 60-70 % for nanocrystalline $(ZrO_2\text{-}3mol\%Y_2O_3)_{80}(Al_2O_3)_{20}$.

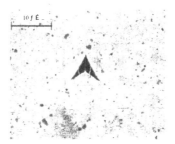

Figure 1. The fully dense specimen of nanocrystalline tetragonal $(ZrO_2-3mol\%Y_2O_3)_{80}(Al_2O_3)_{20}$, as prepared by consolidation of the amorphous powder, with Berkovich indent and its depth profile.

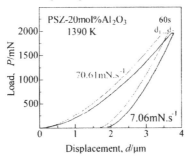

Figure 2. The depth-load curve for fully dense nanocrystalline $(ZrO_2-3mol\%Y_2O_3)_{80}(Al_2O_3)_{20}$ specimen in Berkovich nanoindentation under two different loading rates of 70.6 and 7.1 mN s^{-1}.

Evaluate the yield stress (σ_y) at the room temperature. With the numerical analysis of the finite element method, Larsson and Giannakopoulos present the relationship between load (P) and depth (d) in Berkovich indentation[8]:

$$P = 1.273(\tan24.7°)^{-2}\sigma_y(1+\sigma_{flow}/\sigma_y)(1+\ln E\tan24.7°/3\sigma_y)d^2 \qquad (1)$$

where σ_{flow} is flow stress. Figure 3 shows an analysis of load-depth relation in nanoindentation for the derivation of the local yield stress in consolidated nanocrystalline $(ZrO_2)_{80}(Al_2O_3)_{20}$. When assuming the non-strain hardening ($\sigma_{flow}=\sigma_y$) in an nanocrystalline solid, the prediction based on equation (1) gives a good fit to the experimental result of consolidated $ZrO_2-20mol\%Al_2O_3$ with the crystallite size of 13-17 nm, and the local yield stress, σ_y under constrained compression is deduced at 4.4 GPa: this relatively small value is acceptable for the level of 800 DPN for the Vickers hardness number (Hv)[9]. Then, the σ_y increases to 4.8 GPa with an increased crystallite size of 24 nm for tetragonal $(ZrO_2-3mol\%Y_2O_3)_{80}(Al_2O_3)_{20}$ and tends to increase with increasing loading rate as shown in Fig.2. and so is recognized as the parameter of microplasticity in the nanocrystalline ceramics. It can be generally seen that the hardness decreases, following an ultimate increase, with decreasing grain size (d_o) in nanocrystalline ceramics. Refer to the previous proposed model mechanics for nanocrystalline alloy[10] for a transition of deformation mechanics at the critical crystallite size (d_c). The relationship of the form, $\sigma_y = \sigma_{ya}\{1 + V_f(E_n/E_a - 1)\}$, where σ_{ya} is the yield stress for the amorphous alloy, V_f is the volume fraction of the nanocrystallite, E_n and E_a is Young's modulus for nanocrystal and amorphous phase respectively, may be applicable to nanocrystalline

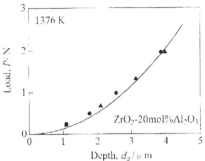

Depth, d_2 / μ m

Figure 3. An analysis of load-depth relation in nanoindentation for the fully dense nanocrystalline $(ZrO_2)_{80}(Al_2O_3)_{20}$ specimen as consolidated at 1376 K.

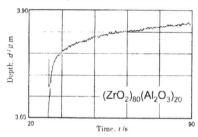

Figure 4. Time-dependent displacement at ambient temperature under constant load in instrumented Berkovich indentation method for fully dense nanocrystalline $(ZrO_2)_{80}(Al_2O_3)_{20}$ specimen.

ceramics by combining with a greatly decreased σ_{ya}, which may be resulted from an increased free volume of the amorphous intercrystal. While, Hall-Petch equation ($\sigma_y = \sigma_{yo} + kd_0^{-1/2}$, σ_{yo} and k is usual meaning) is used to describe the σ_y for nanocrystalline ceramics above the d_c.

Next, provide the phenomenological treatment of a time-dependent displacement under constant load in Berkovich indentation as shown in Figure 4. The plastic strain rate is defined by

$$\dot{\varepsilon} = \partial(d_d/d_t)/\partial t \qquad (2)$$

Combining equation (2) with the relation of the form, $\sigma = P/26.43d^2$ for calculation of true stress under Berkovich indent gives the following equation,

$$\sigma = A\varepsilon^m \qquad (3)$$

where m is the strain rate sensitivity exponent. Figure 5 shows the strain rate in Berkovich nanoindentation versus applied stress for fully dense nanocrystalline $(ZrO_2)_{80}(Al_2O_3)_{20}$. The strain rate sensitivity exponent is derived at 0.014 for nanocrystalline $(ZrO_2)_{80}(Al_2O_3)_{20}$; this small value is in good agreement with predicted[11] and experimental values[6] for non-homogeneous flow in the amorphous alloy. This finding implies that nanocrystalline ceramics undergoes slip deformation under constrained yielding as well as the case of amorphous ceramics[12]. It is interesting to note that the microplasticity related to non-homogeneous non-Newtonian flow gives a rise to various flow events, including the microforging of nanoparticle, and high densification of amorphous powder compact at low temperatures, and is conducive to macroscopic ductilization.

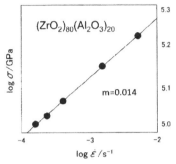

Figure 5. The strain rate in Berkovich nanoindentation versus applied stress for the fully dense nanocrystalline $(ZrO_2)_{80}(Al_2O_3)_{20}$ specimen.

HIGH STRENGTH AND TOUGHNESS

The strength enhanced via nanocrystalline synthesis in monolithic ceramics is evaluated by three-point bend testing. The nanocrystalline $(ZrO_2\square3mol\%Y_2O_3)_{80}(Al_2O_3)_{20}$ plate with a sharp edge, as prepared by mechanical cutting and polishing, shows two fragments in three-point bend testing like a shearing fracture as shown in Figure 5. While the conventional processed ceramics generally exhibits scattering into pieces. The fracture stress (σ_f) in three-point bending is given by

$$\sigma_f = 3FL/2WH^2 \qquad (4)$$

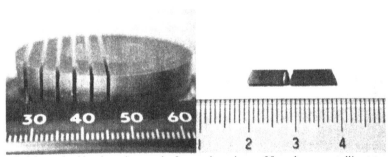

Figure 5. A large sized bend specimen and a fractured specimen of forged nanocrystalline tetragonal $(ZrO_2\text{-}3mol\%Y_2O_3)_{80}(Al_2O_3)_{20}$.

where F is the applied load at fracture and L is the distance between both fulcrums. The flexural strength of monolithic tetragonal $(ZrO_2\text{-}3mol\%Y_2O_3)_{80}(ZrO_2)_{20}$ (d_0=40 nm) does not obey the mass effect as is governed by a weakest link model in a range of the sample volume of this experiment. When taking the aspect ratio (W/H) of the sample width to height as a dimensional parameter, one can see a correlation between flexural strength and W/H for both samples of forged and consolidated tetragonal $(ZrO_2\text{-}3mol\%Y_2O_3)_{80}(Al_2O_3)_{20}$ with d_0=40 nm, having a peak of 2 GPa at $W/H\approx1$ as shown in Figure 6[12]. Therefore, the author tentatively writes this size effect of flexural strength for consolidated nocrystalline ceramics as follows:

$$\sigma_f = F\{W/l\} \tag{5}$$

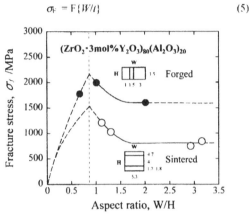

Figure 6. Flexural strength versus aspect ratio for the monolithic tetragonal $(ZrO_2\text{-}3mol\%Y_2O_3)_{80}(Al_2O_3)_{20}$ specimen with the average crystallite size of 40 nm.

Figure 7. Median crack indentation under an applied load of 49 N for as consolidated nanocrystalline $(ZrO_2\text{-}3mol\%Y_2O_3)_{80}(Al_2O_3)_{20}$ specimen with the average crystallite size of 28 nm.

The indentation-microfracture (IM) method is conveniently used to evaluate the fracture toughness (K_{IC}) for consolidated $(ZrO_2\text{-}3mol\%Y_2O_3)_{80}(Al_2O_3)_{20}$ with the crystallite size of 28 nm as shown in Figure 7. The median crack indentation, where the local plastic flow is a dominating factor for the resistance against crack extension, provides the following relationship of the form:

$$P/c_0^{3/2} = K_{IC}/C(E/Hv)^N \tag{6}$$

where C and N are the constants. Figure 8 shows the load (P) versus crack length (c_0) for the full-density sample of nanocrystalline zirconia. For monolithic tetragonal $(ZrO_2\text{-}3mol\%Y_2O_3)_{80}(Al_2O_3)_{20}$ avoiding the formation of monoclinic phase by adding Y_2O_3, the K_{IC} tends to increase with decreasing the crystallite size, and is enhanced to a high level of 30 MPa $m^{0.5}$ via severe compressive forging[13]. Then, thus-obtained fracture toughness has a proportionality to flexural strength in the case of $W/h>2$ for forged and consolidated nanocrystalline $(ZrO_2\text{-}3mol\%Y_2O_3)_{80}(Al_2O_3)_{20}$ as shown in Figure 9. This dependence clearly shows the avoidance of a trade off relationship, which sub-micron sized monolithic ceramics commonly follows, and is so written by

$$K_{IC} = \beta \sigma_F \qquad (7)$$

where β is the constant. In other words, the fracture event in three-point bend testing undergoes an equivalent mechanics underlying median crack extension in IF method for nanocrystalline ceramics.

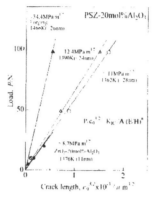

Figure 8. The load versus crack length for full-density nanocrystalline ceramics as prepared by non-equilibrium PM method for the derivation of fracture toughness by IM method.

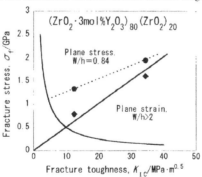

Figure 9. Fracture toughness as evaluated via IM method against flexural fracture strength for nanocrystalline $(ZrO_2\text{-}3mol\%Y_2O_3)_{80}(Al_2O_3)_{20}$, including a trade off relationship.

Consider a model mechanics of fracture in nanocrystalline ceramics. Refer to the critical tensile stress criterion ($\sigma_f = K\sigma_y$, σ_f is a local stress at fracture and K is the plastic constraint factor), a considerably low level of σ_y of monolithic nanocrystalline ceramics leads to a large amount of inhomogeneous plastic flow in a small specimen at the tip of notch or crack, up to a level of microfractural stress increased by the absence of sintering additive, full densification and compressive forging. Consequently, monolithic nanocrystalline ceramics can exhibit enhancements of flexural strength and toughness. Moreover, the plastic constraint at the crack tip is a factor to determine an increase in yield stress at the tip of notch, which leads to a decrease in flexural strength with increasing aspect ratio in a transition from plane stress and plane strain as shown in Fig.6. Note that the high strength and toughness brings good machinability in nanocrystalline ceramics as shown in Fig.5.

HIGH-SPEED SUPERPLASTICITY AT LOW TEMPERATURE

The high-speed superplastic flow is evaluated by compression in thermo-mechanical processing equipped with electric current heating. The cylindrical specimen of monolithic nanocrystalline tetragonal $(ZrO_2\text{-}3mol\%Y_2O_3)_{80}(Al_2O_3)_{20}$ undergoes a continuous free forming up to a relatively high value of $0.75^{[13]}$ for the compressibility (ϕ) avoiding the formation of macroscopic cracks as shown in Figure 10. The superplastic flow in bulk nanocrystalline ceramics.

Figure 10. The superplastic formed nanocrystalline $(ZrO_2\text{-}3mol\%Y_2O_3)_{80}(Al_2O_3)_{20}$ disk, together with as consolidated cylindrical samples prior to compression.

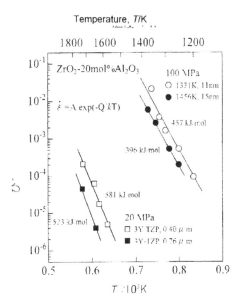

Figure 11. High-speed superplastic flow in compression for nanocrystalline $(ZrO_2)_{80}(Al_2O_3)_{20}$. This figure includes the case of sub-micron sized 3Y-TZP ceramics[14] for comparison.

as prepared by non-equilibrium solid state PM method, is semi-empirically expressed by

$$\varepsilon^{''} = B\sigma(1/d_0)^p \exp(-Q/kT) \qquad (8)$$

where B is the pre-exponential term, m is the strain rate sensitivity exponent, p is the inverse grain size exponent and Q is the apparent activation energy. The nanocrystalline $(ZrO_2)_{80}(Al_2O_3)_{20}$ (11 nm) shows high-speed superplastic flow under the applied pressure of 100 MPa having a strain rate more than 1×10^{-2} s^{-1} at 1358 K during heating as shown in Figure 11. Isochronal method reveals that the temperature necessary for superplastic flow in nanocrystalline $(ZrO_2)_{80}(Al_2O_3)_{20}$ is lower than that of sub-micron sized zirconia (3Y-YZP) [14] by 300-400 K. Then, the isothermal compression in the monolithic nanocrystalline tetragonal $(ZrO_2-3mol\%Y_2O_3)_{80}(Al_2O_3)_{20}$ specimen shows a large variation in true strain rate and true stress at various temperatures without grain growth, which meets the test's condition of phenomenological treatment on the basis of equation (8), as shown in Figure 12.

Under the constancy of d_0 and T, the m here is written by using a relation of the form:

$$m = \partial \log \varepsilon / \partial \log \sigma \qquad (9)$$

A double-logarithm of eqn (8) with the strain rate reduced by the true stress gives

$$\ln(\varepsilon^{''}/\sigma) = \ln B(1/d_0)^p - Q/kT \qquad (10)$$

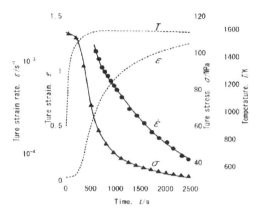

Figure 12. Isothermal compression in nanocrystalline tetragonal $(ZrO_2\text{-}3mol\%Y_2O_3)_{80}(Al_2O_3)_{20}$ in thermal-mechanical testing equipped with electric current heating.

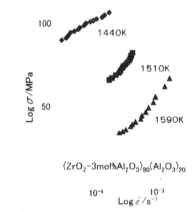

Figure 13. Derivation of the strain rate sensitivity exponent for nanocrystalline tetragonal $(ZrO_2\text{-}3mol\%Y_2O_3)_{80}(Al_2O_3)_{20}$ with the average grain size of 40 nm in isothermal compression.

Figure 13 shows the relationship between true strain rate and true stress according to equation (10) at various temperatures of 1440, 1510 and 1590 K for nanocrystalline tetragonal $(ZrO_2\text{-}3mol\%Y_2O_3)_{80}(Al_2O_3)_{20}$ with d_0=40 nm[15]. The m has a range of the value between 0.2 and 0.7, dependent on strain rate and temperature, showing a non-Newtonian flow. Figure 14 shows

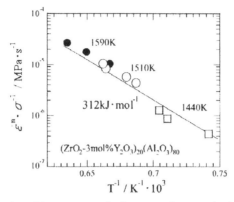

Figure 14. The derivation of the apparent activation energy for superplastic flow in full-density nanocrystalline tetragonal $(ZrO_2\text{-}3mol\%Y_2O_3)_{80}(Al_2O_3)_{20}$.

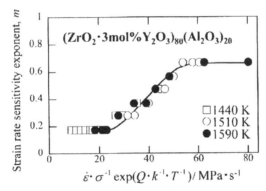

Figure 15. The strain rate sensitivity exponent versus $\varepsilon\,\sigma^{-1}\exp(Q/kT)$ for nanocrystalline tetragonal $(ZrO_2\text{-}3mol\%Y_2O_3)_{80}(Al_2O_3)_{20}$.

the ε^m/σ versus reciprocal temperature for tetragonal $(ZrO_2\text{-}3mol\%Y_2O_3)_{80}(Al_2O_3)_{20}$ with d_0=40 nm. The slop of straight line in this Arrhenius plot permits to obtain the value of 312 kJ mol^{-1} for the level of the activation energy under an electric field for nanocrystalline tetragonal $(ZrO_2\text{-}3mol\%Y_2O_3)_{80}(Al_2O_3)_{20}$. Figure 15 shows the m as a function of the temperature compensated strain rate $\varepsilon\,\sigma^{-1}\exp(Q/kT)$ with the Q of 312 kJ mol^{-1}, known as Zener-Hollman parameter (ZH). reduced by the reciprocal true stress for tetragonal $(ZrO_2\text{-}3mol\%Y_2O_3)_{80}(Al_2O_3)_{20}$ as prepared in the absence of sintering additive. The relationship between the m and $\varepsilon\,\sigma^{-1}\exp(Q/kT)$ is described by the master curve as a shift factor. In order to provide a way to a process window and control for the superplastic flow, thus-obtained dependence here is expressed by

$$m = f[\dot{\varepsilon}\ \sigma^{-1}\exp(Q/kT)] \qquad (11)$$

In the case of tetragonal $(ZrO_2\text{-}3mol\%Y_2O_3)_{80}(Al_2O_3)_{20}$ with $d_o=40$ nm, the m has two constancy of 0.7 and 0.3 for superplastic flow at higher ZH, and the constant value of 0.2 for power law creep at the lowest ZH respectively. On the other hand, the powder consolidation of cubic $(ZrO_2)_{80}(Al_2O_3)_{20}$.

Figure 16. Full-density nanocrystalline $(ZrO_2\text{-}3mol\%Y_2O_3)_{80}(Al_2O_3)_{20}$ sample with both boss extruded in thermal mechanical testing equipped with pulse current heating.

SiC and hydroxyapatite $\{Ca_{10}(PO_4)_6(OH)_2\}$[16] with around 10 nm is formulated by Newtonian flow with $m=1$ combined with an Arrhenius-type equation of the viscosity, $\eta=\eta_o\exp(H_v/kT)\}$ as well as the case of the amorphous phase.

Consider a rate controlling mechanism. For tetragonal $(ZrO_2\text{-}3mol\%Y_2O_3)_{80}(Al_2O_3)_{20}$ with $d_o=40$ nm above the critical crystallite size, a model of grain switching accommodated by diffusional flow especially undergoing grain boundary sliding is appropriate, since the grain elongation and growth do not occur during a low temperature superplastic flow. Now, the Q is about a half value of approximately 600 kJ mol^{-1} for diffusion creep in the $ZrO_2\text{-}Al_2O_3$ system and so supports the diffusional flow inside a grain boundary with a relatively large free volume. Then, it is interesting to compare with a theoretical model such as solution-precipitation creep in glass ceramics having grain boundary with a thin liquid film. On the other hand, for the nanocrystalline ceramics having approximately 12 nm below the critical crystallite size and a large amorphous volume, a core mantle mechanism is thought to be responsible for Newtonian viscous flow.

Figure 16 shows the full-density nanocrystalline $(ZrO_2\text{-}3mol\%Y_2O_3)_{80}(Al_2O_3)_{20}$ disk with the boss at both side as extruded at a relatively low temperature by using specially designed graphite rod in thermal-mechanical testing equipped with pulse electric current heating. One looks elsewhere for the working map for net-shape-forming via superplastic extrusion in nanocrystalline materials[17]. The high-strain-rate superplastic forming can be generalized for nanocrystalline oxide and covalent ceramics and has the potential for net shape control in widespread applications.

CONCLUSIONS

This article briefly describes the phenomenology of flow and fracture occurred in nanocrystalline ceramics, namely with the average grain size below 50 nm, as prepared by pulse electric current consolidation of the mechanically synthesized amorphous powder. The material testing described here can be elegantly polished for nanocrystalline ceramics by introducing the progressed analytical methods of nanoindentation and compression, together with the evaluation by fracture mechanics and plasticity theory. The research of nanocrystalline ceramics is still growing in various nanoprocessing, including the recent interesting works on nanoscale composites with grain size of around 90 nm at the minimum[18] in solid state P/M method followed by some researchers. Since now, a model nanomechanics study is so expected to provide a way of functionalization of nanocrystalline and amorphous ceramics by combining with high resolution electron microscopy and computer simulation based on molecular dynamics.

REFERENCES

[1]H. Kimura: *Materials Integration*, **12**, 19-26 (1999). *Japanese patent* 3799459 (filed on 1997.11.11), 297-301 (2007).

[2]H. Kimura: *Advances in Powder Metallurgy & Particulate Materials* (MPIF), **12**, 55-61 (1999).

[3]H. Kimura and K. Hongo: *J. Jpn Inst. Metals*, **63**, 649-655 (1999).

[4]H. Kimura: *J. Metastable and Nanocryst. Mater.*, **15-16**, 591-598 (2003).

[5]H. Kimura and Y. Fujimoto: *J. Jpn Soc. Powder Powder Metallurgy*, **46**, 1274-1283 (1999).

[6]H. Kimura and T. Masumoto: *Amorphous Metallic Alloys* (Butterworths, London), 187-230 (1983).

[7]H. Kimura: *Iron and Steel Institute of Japan (ISIJ)*, **72**, 1498-1506 (1986).

[8]P.-L. Larsspm and A.E. Giannakopoulos: *Int. J. Solids Structures*, **33**, 221-248 (1996).

[9]H. Kimura and K. Hongo: *Materials Transaction*, **47**, 1374-1379 (2006).

[10]H. Kimura and A. Hachinohe: *Advanced Particulate Materials & Processes* (MPIF), 153-160 (1997).

[11]F. Spaepen: *Acta Met.*, **25**, 407-415 (1977).

[12]A.S. Gandhi and V. Jayaram: *Acta Mater.*, **51**(2003)1641-1649.

[13]C. M. Aji: Master thesis, National Defense Academy Japan, (2002).

[14]M. Nauer and C. Carry: *Euro-Ceramics*(Elsevier, London UK), **3**, 3323 (1989).

[15]T. Uchino: Master thesis, National Defense Academy Japan, (2004).

[16]H. Kimura, K. Hanada and K. Ishigane: *Materials Science Forum*, **502**, 211-216 (2005).

[17]T. Uchino and H. Kimura: *Materials Processing for Properties and Performance* (MP³), **2**, 93-99 (2004).

[18]K. Morita and K. Hiraga: *Materials Integration*, **20**, 60-65 (2007).

ACKNOWLEDGEMENTS

Thanks are due to Mr. C. M. Aji and Mr. T. Uchino, graduate students of National Defense Academy Japan for the technical assistance during this investigation.

PROPERTIES OF NANOSTRUCTURED CARBON NITRIDE FILMS FOR SEMICONDUCTOR PROCESS APPLICATIONS

Jigong Lee, Choongwon Chang, Junam Kim and Sung Pil Lee
Department of Electronic Engineering, Kyungnam University
Masan, Kyungnam, Korea

ABSTRACT

The crystalline properties of nanostructured carbon nitride films are analyzed by XRD, SEM, XPS, EDS and FTIR for semiconductor process applications. The thickness and growth rate of CN_x films with different growth conditions are reported and an empirical model is proposed for the thickness calculation to find growth conditions. The deposited carbon nitride has β-C_3N_4 and lonsdaleite peaks, uniform nanostructured surface which has grain size of about 50 nm. . The chemical formula of the deposited carbon nitride film is roughly expressed as C_7N_4 and C_3N.

INTRODUCTION

The first researches of carbon nitride (CN_x) were focused only on synthesizing the stoichiometric -C_3N_4 phase which was suggested as a new super hard material that might have similar or superior hardness and bulk modulus of a diamond by Liu and Cohen [1], but most of the researchers have not found the dream materials. The structure of β-C_3N_4 consists of buckled layers stacked in an AAA… sequence [2].

The unit cell is hexagonal and contains two formula units (14 atoms) with local order such that C atoms occupy slightly distorted tetrahedral sites while N atoms sit in nearly planar triply coordinated sites. This structure can be thought of as a complex network of CN_4 tetrahedra that are linked at the corners. The atomic coordination suggests sp^3 hybrids on the C atoms and sp^2 hybrids on the N atoms. The hexagonal unit cell and the planer coordination of the N sites raise the possibility that this structure may exhibit anisotropic elastic properties. It is now clear that the production of crystalline β-C_3N_4 has been the goal of much research. There have been many reports of amorphous carbon nitride films of uncertain composition but there have only been a few observations of crystalline β-C_3N_4 produced by reactive sputtering, laser ablation and hot filament CVD [3-5]. The films reported have been discontinuous with isolated crystals [5] or showing only a few isolated grains in an amorphous matrix [4] and in most cases high substrate temperatures (600 – 950 °C) were required. According to a report of R. C. DeVries [6], although over 400 papers have been published on crystalline C_3N_4 by 1997, no new super hard materials have come out over a decade. One of the most significant problems degrading the quality of carbon nitride films is existence of N-H and C-H bonds mostly found in low temperature sputtering systems. The possibility of these reactions with hydroxyl group of carbon nitride films, caused by a hydrogen attack, was suggested in our previous reports and proved that these undesired effect could be applied for fabricating semiconductor sensors [7].

Carbon nitride has several advantages for semiconductor process applications as follows:

1. It could be nanostructured thin film with sub-micron orders.

2. Growth methods for the film, using sputtering or chemical vapor deposition are well developed and easily applicable to the standard semiconductor process.

3. It has high thermal and chemical stability, meaning a long lifetime and long-term stability and this can make it possible to apply to harsh conditions.

4. It has good adhesion to silicon wafer, silicon oxide and other substrates.

5. It has good uniformity on the whole substrate and low roughness.

In this paper, the crystalline properties of nanostructured carbon nitride films are analyzed by XRD (X-ray diffractometer), SEM (scanning electron microscopy), XPS (X-ray photoelectron spectroscopy), EDS (energy dispersive X-ray spectrometer) and FTIR (Fourier transform infrared spectrometer) for semiconductor process applications.

EXPERIMENTAL

The films were deposited by reactive RF magnetron sputtering. The use of a Nd-Fe-B magnet ensures a very high flux density in between the substrate and target region[6]. A high-purity 99.997% graphite planar target was mounted at a distance of 6 cm from the substrate and different N_2/Ar gas mixtures were used as the sputtering gases to vary the film compositions. The gas purity was 99.999% in each case. Prior to deposition, the substrates were cleaned with acetone and methanol in an ultrasonic bath. All the substrates were cleaned by an RF glow discharge of Ar for 5 min to remove any residual contaminants. Flow rate ratio of N_2/Ar for film formation was varied with 0/10, 3/7, 5/5, 7/3, and 10/0. Other deposition conditions of the carbon nitride films were fixed as follows; RF power: 200 W, substrate bias: -60 V (DC), temperature: 200 °C, deposition time: 60 min, working pressure: 5 mTorr [8].

RESULTS AND DISCUSSIONS

Alpha-step profilometer is one of the common methods chosen for the thickness measurement. To make steps for measurement, test patterns of the films are prepared usually by the lift-off technique. The thickness and growth rate of CN_x films with different growth conditions are reported and an empirical model of the thickness is included for the thickness calculation to find the growth conditions for aimed thickness. When we pursuit the growth of crystalline carbon nitride, short distance between target and substrate, high RF power, and high temperature of the chamber are needed to produce higher reaction power. However, on the contrary, lower power and longer distance are better for uniformity of the films and reproducibility. The relationships between the thickness and some growth conditions will be a fundamental factor of the physical parameter extractions.

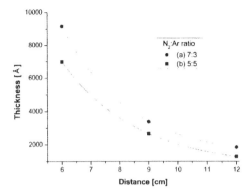

Figure 1. The thickness of CN$_x$ films as a function of the distance between target and substrate.

Fig. 1 shows the thickness changes as a function of the distance between the target and the substrate. The two sample of 7:3 (Fig. 1 (a)) and 5:5 (Fig. 5.1 (b)) N$_2$:Ar ratio are prepared with 100 W RF power at 150 ⁻ chamber temperature for 30 min. The thickness is drop off by inverse square law. The thickness of the sample (a) and (b) are 1854 ⊓ and 1285 ⁻, respectively. The thickness of the sample with 7:3 N$_2$:Ar ratio is 30 % higher than that of sample with 5:5 N$_2$:Ar ratio. The growth rate can be calculated as 3708 ⊓/hr (sample (a)) and 2570 ⁻/hr (sample (b)). From the result, the imperial equation can be obtained by the following expression

$$t = \frac{k \times m \times T}{(d - 1.2)^2} [A] \qquad (1)$$

where, k is a proportional constant [⁻²], m is the growth rate [⊓/min], and T is deposition time [min]. The asymptote is adjusted at 1.2 in the denominator to compensate an error in the real system. For the sample (a) in the Fig. 2, the parameters, T = 30 [min], m = 226.46 [⁻/min], and k = 30.9105 [⁻²] are given. From this relationship, when the distance is given, the thickness of the films can be calculated as a function of deposition time.

Fig. 2 illustrates the relationship between the thickness of CN$_x$ films and deposition time with condition of annealing at 450 °C for 30 min. The films are deposited at 120 °C with 70 % nitrogen incorporation. The thickness of the films is proportional to the deposition time from 633 ~ 3565 ⊓ with a gradient, m = 58.64 [⁻/min]. After annealing at 450 °C, the thickness decreases by about 40 %. Growth rates of the films are 0.356 /hr (as-deposited, Fig. 2(a)) and 0.21 /hr (anealed at 450 °C, Fig. 2 (b)), respectively.

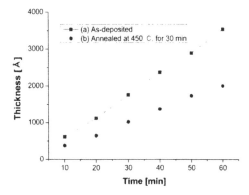

Figure 2. The thickness of CN_x films as a function of deposition time: (a) as-deposited and (b) after annealing at 450 °C for 30 min.

To examine the crystal structure of carbon nitride films. XRD analysis is performed with a X'Pert APD system (Philips. Netherland) using CuKα radiation ($\lambda = 1.54$) scanning from 10 to 120° of 2θ. The sample is prepared on the p-type silicon wafer (100) by reactive RF magnetron sputtering with 160 W RF power for 1 hour 30 min. The ratio of N_2/Ar is fixed at 7:3 and controlled by MFC. Fig. 4 shows the XRD spectra of the as-deposited carbon nitride films with the different DC biases: (a) -90, (b) -60, and (c) -30 V. All samples in Fig. 3 show β-C_3N_4 (200) and lonsdaleite (102) peaks. i.e. 32.533°(PDF 50-1512) and 61.852°(PDF 19-0268). respectively. The strong peak at 69.19°is the Si (400) peak with d-spacing. 1.3569⁻. There are no graphite peaks in the films. These results reveal that the complete chemical reaction is occurred during the film formation. As shown in Fig. 3. as the absolute value of the DC bias increases from −30 to −90 V. the β-C_3N_4 (200) peak intensity increases. However. we do not observe a considerable change in the lonsdaleite (102) peak intensity according to DC bias. It seems that applied negative bias etches weak bonded C-N sites and increases crystallization. However. in high negative bias over -120 V. the plasma is very unstable and intermittently. undesirable arc is occurred on the target.

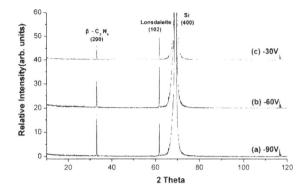

Figure 3. XRD patterns of CN_x films on silicon wafer at 160 W RF power for 1 hour 30 min with different DC bias: (a) -90 V, (b) -60 V, and (c) -30 V.

Fig. 4 shows SEM photographs of the surfaces and the cross-sections of as-deposited CNx films onto a silicon substrate with 0, 30, 50, and 70% nitrogen incorporation. The RF power is 300 W, the DC bias is -80 V, the deposition time is 20 min, and the heating temperature is 150 °C, respectively. As shown in Fig. 4, the deposited films with over 10% nitrogen incorporation have small grains with a hexagonal structure. Fig. 4(c) in case of 70% nitrogen shows the biggest grain size ca. 50 nm. The cross-sections of the films show a clear border between CNx films and Si-wafer, and the thickness of deposited films with 70% nitrogen is about 346 nm. When the nitrogen incorporation is 70%, growth rate is highest and grain size is biggest. Even though the factors of RF power, negative DC bias, and chamber temperature increase the crystallization of films, weak C-N bond in CNx film is needed for semiconductor processes. The deposition conditions of low power, no DC bias, and low chamber temperature are adopted intentionally to increase weak C-N bonding site and to avoid the damage of photo resist due to a high sputtering energy. Over 70% nitrogen sputtering gas ratio, growth rate is rather decreased.

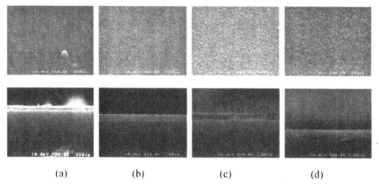

(a)   (b)   (c)   (d)

Figure 4. SEM photographs of cross-section and surface of CNx films as different N_2 ratios: (a) N_2 0 %. (b) N_2 10 %, (c) N_2 50 % and (d) N_2 70 %.

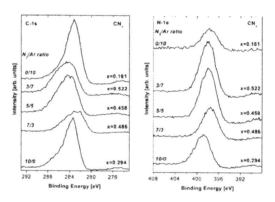

Figure 5. The X-ray photoelectron spectrum of C 1s(a) and N 1s(b) electrons of CN_x films.

Fig. 5 shows the XPS spectrum of C 1s and N 1s electrons of CN_x films as a function of nitrogen ratio ($N_2/(N_2+Ar)$). The films are deposited with 200 W RF power at 200 °C for 1 hour. According to the magnified peaks of C 1s and N 1s in Fig. 5, the C 1s peaks are shifted to the right about 2 eV from the β-C_3N_4 peak. 286.2 eV and 286.8 eV. However, the N 1s peaks are very close to β-C_3N_4 peaks. 398.6 eV and 398.6 eV. It can be considered that the deposited films are rich in C-C bond but the theoretical β-C_3N_4 film is not. Atomic concentrations of C and N can be calculated from the area of the XPS data in Fig. 5. When the nitrogen ratios are 30, 50 and 70 %, the nitrogen incorporations are 34.3, 31.4 and 32.7 %, respectively. The chemical formula can be expressed roughly as C_2N or C_5N_2. From these results, it can be suggested that the film is amorphous carbon-rich phase with a partial β-C_3N_4 phase.

The composition ratio of deposited films is investigated by EDS which measures deeper surface than XPS or AES. The films are deposited with 200 W RF power at 200 °C for 1 hour. Fig. 6 shows the normalized EDS peaks of the samples and the composition ratio that is extracted from the result. The highest value and lowest value of carbon/nitrogen ratio in CNx films are 64.93/4.7 at 0% N_2 sputtering gas ratio and 51.97/36.11 at 50% N_2 sputtering gas ratio, respectively. When the chemical formula is represented as $C_{1-x}N_x$, the range of x is 0.25 to 0.36. Therefore, it can be roughly expressed as C_7N_4 and C_3N. These results are similar to the XPS and AES results (roughly C_2N). The detail of component is shown in table 1. As the N_2 sputtering gas ratio increases, nitrogen incorporation in the film also increases under the condition that the nitrogen gas ratio is lower than 50%. However, when N_2 ratio is higher than 50%, the nitrogen incorporation of the film is even decreased. This is probably due to increased sputtering rate of C species from the target as the steady state N concentration increases on the target surface. Energetic or neutral Ar species probably enhance the mobility of nitrogen species and increase N sticking at the growth surface. There is indeed an effect of Ar mixture on the nitrogen incorporation in the film. When N_2/Ar ratio is 30 ~ 50%, sputtering rate of carbon and nitrogen sticking effect are maximum at the growth surface. If Ar increases further in the sputtering gas mixture, two things can happen, (1) the sputtering rate of carbon increases with respect to ionized nitrogen species, thus the film contains less nitrogen; (2) chemically enhanced preferential sputtering of nitrogen from the film surface can occur by the energetic Ar species. Increase in Ar gas in the sputter gas mixture may also disrupt the film structure. The momentum of the Ar^+ ions is higher than that of either N_2^+ or N^+ ions, and then the increased momentum transfer into the growing film causes disruption of the bonding structure of the film producing amorphous material. We can find that if N_2 increases more than 70% in the sputtering gas mixture, sputtering yield and N sticking effect are decreased due to deficiency of energetic Ar species. It can be seen that with higher than 70% N_2 sputtered film, nitrogen incorporation is lower than the 50% N_2 gas sputtered film. A small amount of oxygen (less than 6.23 atomic%) is due to contamination from the chamber wall, and Si peak is caused by the silicon wafer substrate.

Table 1. The atomic ratio of CN_x films by EDS.

N_2 ratio (%) Element	0	30	50	70	100
C	64.93	55.14	51.97	71.68	69.41
N	4.7	31.59	36.11	23.63	24.44
O	2.64	4.60	6.23	3.67	4.81
Si	25.9	8.66	5.69	1.03	1.34

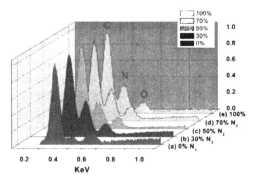

Figure 6. Normalized EDS spectrum of CN_x films with different N_2 ratio.

Fig. 7 shows the substrate affection on the chemical bonding state of CN_x films. The films grown on alumina and silicon nitride have relatively strong crystalline structure because the hexagonal structure of alumina and zinc blende structure of silicon nitride serve as a seed of the crystallization and increase the D/G ratio. Silicon nitride also have strong absorbtion peaks of around 3300 cm^{-1} attributed to >N-H and -NH$_2$. It is because that the nitrogen radical in the silicon nitride help increase of the nitrogen site of the films providing hydrogen bonds.

To find the influence of the water vapor adsorption, two samples were prepared. The humidified CN_x film put in the chamber over 90 %RH for 24 hour and the dried sample put in the dehydrated chamber by N_2 purge for 24 hour. The spectrum of humidified CN_x films (Fig. 8 (a)) and the spectrum of the dried CN_x films (Fig. 8 (b)) have both Si-N (940 cm^{-1}), D (1270-1410 cm^{-1}), G (1520-1590 cm^{-1}), C=N (1600-1700 cm^{-1}), C≡N (2200 cm^{-1}), nitrile group (2367 cm^{-1}) and C-H (2970 cm^{-1}). However, some peaks, such as ≡C-H (3300 cm^{-1}), -OH (3400 cm^{-1}), >N-H and -NH$_2$ (3200 - 3380 cm^{-1}) and H$_2$O (3500 cm^{-1}), are dominant in the humidified CN_x films, and are remarkably reduced in dried CN_x films. It demonstrates that CN_x film reacts easily with hydrogen and/or hydroxyl group. We can see the peak shift due to the water vapor adsorption. The spectrum of Si substrate (Fig. 8 (c)) is for reference.

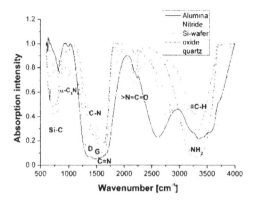

Figure 7. FTIR spectra of CN$_x$ films on different substrate.

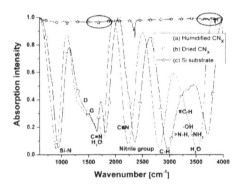

Figure 8. FTIR spectra of (a) humidified CN$_x$ film, (b) dried CN$_x$ film, and (c) Si substrate.

CONCLUSIONS

Carbon nitride has several advantages that could be nanostructured thin film with sub-micron orders and easily applicable to the standard semiconductor process. The physical and chemical properties of nanostructured carbon nitride films were analyzed by XRD, SEM, XPS, EDS and FTIR. From the proposed imperial model, when the distance is given, the thickness of the films can be calculated as a function of deposition time. The deposited films show β-C$_3$N$_4$ and lonsdaleite peaks in XRD, and completely chemical reaction is occurred during the film formation due to no graphite peaks in the films. We can find that the hydrogen attack in carbon nitride film could easily break or change the C≡N and C=N bonds to form C-H and N-H bonds from FTIR.

ACKNOWLEDGEMENT

This research was financially supported by the Ministry of Commerce and Energy (MOCIE), and Korean Industrial Technology Foundation (KOTEF) through the Human Resource Training Project for Regional Innovation.

REFERENCES

[1] A. Y. Liu and M. L. Cohen, Science, 245, 841 (1989).

[2] O. Borgen and H. M. Seip, Acta Chem. Scand, 15, 1789 (1961).

[3] S. P. Lee and J. B. Kang, Microchemical J., 70, 239 (2001).

[4] H. Han, B. J. Feldman, Solid State Commun., 65, 921 (1988).

[5] P. Gonzalez, R. Soto, E. G. Parada, X. Redondas, S. Chiussi, J. Serra, J. Pou, B. Leon and M. Perez-Amor, Appl. Surf. Sci., 109-110, 380 (1997).

[6] R. C. DeVries, Mater. Res. Innovations, 1, 161 (1997).

[7] J. G. Lee and S. P. Lee, J. Korean Phys. Soc. 45, 619 (2004).

[8] J. G. Lee and S. P. Lee, *Sensors and actuators B* 108, 450 (2005).

APPLYING NICKEL NANOLAYER COATING ONTO BB4BC PARTICLES FOR PROCESSING IMPROVEMENT

Xiaojing Zhu, Kathy Lu, Hongying Dong, Chris Glomb, Elizabeth Logan
Department of Material Science and Engineering, Virginia Polytechnic Institute and State University
Blacksburg, VA 24061
USA

Karthik Nagarathnam
Utron, Inc.
Manassas, VA 20109
USA

ABSTRACT
    This research is based on our previous work of applying Ni nanolayer onto the B_4C particle surfaces by an electroless coating technique. The effects of different chemicals used in the electroless coating were studied (the amounts of the activation agent ($PdCl_2$), and the complexing agent ($C_2H_8N_2$), and the addition rate of the reducing agent ($NaBH_4$)). SEM and XPS are employed to characterize the coated B_4C surfaces. The Ni-B nanolayer formation is strongly dependent on the activation sites and the rate that Ni^0 can be reduced. Fundamental Ni-B coating processes and morphological changes on the B_4C particle surfaces are analyzed. Combustion driven compaction is used to produce high density B_4C compacts from Ni-B nanolayer coated B_4C particles.

1. INTRODUCTION
    B_4C is an important non-metallic hard material (Vickers hardness 2900-3900 kg/mm^2) with a high melting point (2450°C). B_4C's unique properties such as its hardness, good chemical resistance, low density (2.52 g/cm^3), and neutron absorption properties make B_4C an important material in a variety of technical applications. These include the machining of diamond tools, the creation of hot-pressed shot blast nozzles, ceramic tooling dies, armors, and shielding, control rod, and shut down pellets in nuclear power plants.
    Despite these attractive properties, it is difficult to process B_4C because achieving high density B_4C components has been a challenge. B_4C particles are extremely hard and do not deform under typical compaction pressure. This is especially true if the surface of the particles is not smooth and the particles are not regular in shape. Under these circumstances, it will be fairly difficult for the particles to move with respect to each other and rearrange under the compacting pressure. Also, because of the high melting point (2450°C) and the low boron and carbon diffusion mobility (a consequence of the covalent bonding), it is difficult to sinter B_4C. Pure B_4C cannot be sintered to densities higher than 75-80% of the theoretical density even at temperatures higher than 2300°C. All these factors make B_4C processing inefficient and expensive.
    In order to make the realistically possible density closer to the theoretical density, there has been a persistent effort in using additives to facilitate the compaction and sintering processes of B_4C. The most typical approach is to mix B_4C with some low melting materials such as Al or Ni to achieve densification.[1] Also additives such as TiB_2[2,3] and TiO_2[4] etc. have been added into the B_4C matrix in the pressureless sintering of B_4C. Although some researchers have taken this approach, this is not always a feasible solution for a mechanically demanding application because the second phase formed always degrades the mechanical performance.

In this context. if a thin metallic layer can be incorporated onto the B$_4$C particle surfaces, it should greatly facilitate the rearrangement of the B$_4$C particles during compaction and the B$_4$C species diffusion during sintering. When the amount of the metallic layer is controlled at a very low level and distributed homogeneously in the B$_4$C matrix, it should introduce new functionalities such as electrical conductivity and magnetism without considerable performance degradation. This makes B$_4$C suitable for more applications. Electroless coating is a very promising technique to provide such a metallic layer on the B$_4$C particle surfaces. Nickel electroless coating has been widely studied because of its unique properties such as good ductility. lubricity. and excellent electrical properties. If metallic nickel can be incorporated into B$_4$C using an electroless coating technique. a uniform thin layer should then form on the B$_4$C surfaces. This coating is not only going to improve the processability of B$_4$C, but also create greater functionality. Electroless nickel coating is being increasingly used to coat metallic layers onto particles, fibers, or even tubes. For example, nickel has been coated onto carbon fibers and SiC particles by an electroless coating technique. [5][6] Although hypophosphate-reduced nickel coating has received widespread acceptance, attention has shifted towards borohydride-reduced nickel deposits in recent years. Apart from the uniformity, electroless nickel-boron coating can produce extremely hard deposition of nickel boride. which in turn provides incredible wear and abrasion resistance. [7] The presence of Ni-B coating is expected to benefit the B$_4$C sintering in this work since it not only contains a low-melting material. but also possesses great mechanical properties and won't compromise the bulk B$_4$C mechanical properties very much. Sodium borohydride (NaBH$_4$) is one of the most powerful reducing agents for electroless nickel boron coating and is chosen for the reducing agent in this work.

In our prior work, we studied electroless coating of Ni-B nanolayer onto B$_4$C particles. [8] Using NiSO$_4$ as a Ni^{2-} source. SnCl$_2$ as a sensitizing agent. PdCl$_2$ as an activation agent, C$_2$H$_8$N$_2$ as a complexing agent, and NaBH$_4$ as a reducing agent. Ni-B nanolayers of different thicknesses were successfully coated onto B$_4$C particles. The Ni-B coating thickness can be adjusted by the Ni^{2-}:B$_4$C molar ratio. In this study, factors that influence the Ni-B nanolayer are investigated. The morphological and chemical compositional differences under different conditions are discussed. Also, primary results on forming high density compacts using the combustion driven compaction technique will be presented.

2. EXPERIMENTAL PROCEDURE

B$_4$C particle size distribution before and after separation are shown in Figure 1. The particle size of B$_4$C particles (H. C. Starck. Inc.. Newton. MA) was measured using a laser light scattering analyzer (Horiba. LA-700. Irvine. CA). The particle size distribution of the as-received particles was bimodal as shown in Figure 1 (a). Therefore centrifugal experiments are necessary to obtain monodispersed particles. A centrifugal technique has been proven to be effective to separate B$_4$C particles. As shown in Figure 1 (b). the tiny sized B$_4$C particles are successfully removed. B$_4$C particles used in this work have a unimodal distribution with the peak value of 2.27 µm.

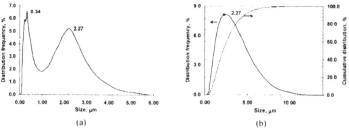

Figure 1. B$_4$C particle size distribution. (a) before separation. (b) after separation.

Figure 2 shows the experimental procedure of electroless nickel coating onto B$_4$C particles. Before the electroless coating, the B$_4$C particles were sensitized using SnCl$_2$ and activated using PdCl$_2$. The surface sensitization was carried out by adding B$_4$C particles into a SnCl$_2$·2H$_2$O (> 98%, Fisher Scientific, Fair Lawn, NJ) and HCl (36~38%, EMD, Gibbstown, NJ) solution (0.07 M SnCl$_2$·2H$_2$O, 40 ml/L HCl). The suspension was sonicated for 5 min followed by magnetic stirring for another 5 min at room temperature. The Sn^{2+} sensitized B$_4$C particles were thoroughly washed with de-ionized water and transferred to a PdCl$_2$ (> 99%, Fisher Scientific, Fair Lawn, NJ) and HCl solution (0.0042 M PdCl$_2$, 40 ml/L HCl) for activation. After mixing at room temperature for another 10 min in the same way as that in the sensitizing step, the activated B$_4$C particles were again thoroughly washed with de-ionized water and then introduced into the electroless coating bath. NiSO$_4$·6H$_2$O (99.0%, Fisher Scientific, Fair Lawn, NJ) was used as the Ni^{2+} source, which was complexed prior to pH adjustment. The complexing agent used in this work was ethylenediamine (C$_2$H$_8$N$_2$, Fisher Scientific, Fair Lawn, NJ).

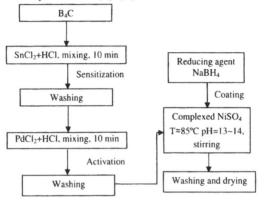

Figure 2. Flow chart of the electroless nickel coating onto separated B$_4$C particles.

The electroless coating temperature was kept at 85±2°C. The pH value of the coating bath was adjusted between 12 and 14 with 10 M sodium hydroxide (Fisher Scientific, Fair Lawn, NJ). Unlike the conventional electroless plating in which the reducing agent was one ingredient of the plating solution, in this work the reducing agent NaBH$_4$ (Fisher Scientific, Fair Lawn, NJ) was added into the coating bath drop-wise. As mentioned in the introduction, the effect of different factors on the nanolayer coating was studied. These factors include the amount of PdCl$_2$ with respect to the amount of B$_4$C in the activation step, the amount of complexing agent with respect to the amount of nickel, and the rate that NaBH$_4$ is added into the coating bath. The conditions studied are summarized in Table I.

Table I. Conditions Studied in Electroless Coating Procedure*

Chemical composition (molar ratio)					NaBH$_4$ addition rate
B$_4$C:Sn^{2+}	B$_4$C:Pd^{2+}	Ni:C$_2$H$_8$N$_2$	Ni:NaBH$_4$	Ni:B$_4$C	All at one time
1:0.7	1:0.04 1:0.01 1:0.005 1:0.001	1:4.5 1:6 1:9	1:1	0.3354	10 drops/min 1 drop/min

A field emission SEM equipped with an EDS (LEO 1550. Carl Zeiss Microlmaging. Inc. Thornwood. NY) was used to characterize the surface morphology of the Ni-B nanolayers. XPS (Perkin Elmer 5400, Minneapolis, MN) was employed to characterize the composition of the Ni-B nanolayers.

Combustion driven compaction of Ni-B nanolayer coated B$_4$C particles was conducted at Utron. Inc. A compaction pressure of 2.63 GPa was applied. The sample diameter was 25.4 mm and the thickness was 1.4 mm. The strength of the green samples was evaluated using equibiaxial flexural strength testing. The apparent density of the samples was determined by measuring the weight and volume of the compacts.

3. RESULTS AND DISCUSSIONS

3.1. Activation Agent Effect

Activation agent PdCl$_2$ functions in the following way. Pd^{2+} oxidizes Sn^{2+} to Sn^{4+} and coverts itself into Pd^0 on the B$_4$C surfaces during the activation step. Pd^0 atoms act as catalytic centers in the initial stage of the electroless coating of the B$_4$C particle surfaces. The reduced Ni atoms then act as the autocatalytic centers for further Ni^{2+} reduction. Based on the atomic radius of Pd^0 and the specific surface area of B$_4$C particles. each Pd^0 will cover about 10^{-19} m^2 surface area of B$_4$C particles. For a monolayer Pd^0 surface coverage on B$_4$C particles. this corresponds to a B$_4$C:Pd^{2+} molar ratio of 1:0.01. Our prior result of 1:0.04 B$_4$C:Pd^{2+} ratio indicated that Pd^{2+} is excessive. consistent with the study by Brandow et al.[9] To evaluate the B$_4$C:Pd^{2+} ratio effect on the Ni-B nanolayer morphology. the B$_4$C:Pd^{2+} molar ratio of 1:0.04, 1:0.01. 1:0.005. and 1:0.001 are studied at Ni:C$_2$H$_8$N$_2$ ratio of 1:6 and NaHB$_4$ addition rate of 10 drops/min.

Figure 3 presents SEM images of the bare and coated B$_4$C surfaces activated using different PdCl$_2$ amounts. First by comparing Figure 3 (a) with other images in Figure 3. the comparison clearly shows that a layer with mesh structure has been successfully coated onto the B$_4$C particle surfaces uniformly. Further examination of Figure 3 tells that. when the B$_4$C:Pd^{2+} molar ratio is 1:0.04. a rough coating layer with large Ni-B nodules (bright spots) is observed. When the B$_4$C:Pd^{2+} ratio is decreased to 1:0.01 (Pd^0 monolayer coverage on the B$_4$C surfaces). the coating roughness remains almost the same. The mesh-like nanostructures among the Ni-B nodules can be easily seen. This means high PdCl$_2$ amount can over-activate the B$_4$C particle surfaces and cause the Ni-B nanolayer to grow too fast. Also. ideal Pd^0 monolayer packing may not be possible and some Pd^0 atoms likely exist as clusters at certain locations. The Ni-B nodules will thus quickly form at the concentrated Pd^0 locations. initiating the formation of larger sized nodules. When the electroless coating process continues under such condition. particles will form on the to-be-coated surfaces, as seen in many studies reported. [5][6] Also. this result is close to Brandow et al.'s work in which a Pd^0 surface coverage of as low as 20% is declared to be enough to initiate complete and homogenous coating.[9] When the B$_4$C:Pd^{2+} molar ratio is decreased to 1:0.005. dense Ni-B coating is obtained and the Ni-B nodule size decreases substantially. This means that fewer activation sites on the B$_4$C particles are more conducive for the Ni-B nanolayer growth while suppressing excessive Ni-B nodule growth. For the studied B$_4$C particles. the Pd^0 concentration on the particles should be half of the Pd^0 monolayer surface coverage. When the B$_4$C:Pd^{2+} molar ratio is further decreased to 1:0.001. the Ni-B nanolayer becomes rough again and the size of the Ni-B nodules increases, even though still much smaller than those of the first two B$_4$C:Pd^{2+} ratios. This surface roughening is likely because the diffusion distance needed for the newly reduced Ni to reach the growing nanolayer edge is too long. For uniform and dense Ni-B nanolayer coating. a B$_4$C:Pd^{2+} ratio of 1:0.005 proves to be the optimal activation condition.

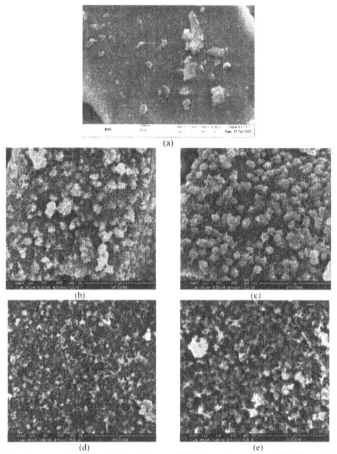

Figure 3. SEM images of (a) bare B$_4$C surface, and Ni-B nanolayers coated B$_4$C particle surfaces activated with a different molar ratio of B$_4$C:Pd^{2+}: (b) 1:0.04, (c) 1:0.01, (d) 1:0.005, and (e) 1:0.001.

The difference in the Ni-B nanolayer morphologies shows that there is an optimal PdCl$_2$ amount in order to achieve the desired nanolayer and avoid particle formation in the coating. Pd^0 concentration affects the Ni-B growth initiation sites, the Ni-B deposition rate, and subsequently the Ni-B nanolayer morphology. When too much PdCl$_2$ is used for B$_4$C surface activation, Ni and B will attach to the B$_4$C surfaces at a high rate. However, the nanolayer growth mode requires the Ni and B species to diffuse to the fresh layer edge. Since some Ni and B will not have enough time to diffuse long distance and attach to the growing layer edge before the arrival of the more reduced species, these diffusing species will adsorb on the top of the previously deposited Ni-B layer. This leads to island-like nodule growth, as seen in Figure 3.

Figure 4 shows the XPS spectra of the original, and Ni-B nanolayer coated B$_4$C particles, activated using different amount of the activation agent PdCl$_2$. The peak rising at around 187 eV represents the B-C bond, while the peak at around 191 eV represents the B-O bond. In the original B$_4$C only the B-C bond is detected. While in the spectra of all coated samples, the B-O peak dominates the XPS spectra. As discussed before, B$_2$O$_3$ forms during the Ni-B nanolayer formation process.[8] The existence of B-O peaks means that the B$_4$C particles are mostly covered with Ni-B nanolayers. The difference among the coated samples is mainly on the B-C peak. When the B$_4$C:Pd^{2-} ratio is 1:0.005, the intensity of B-C peak is the lowest. A very weak B-C bond from the B$_4$C particle surfaces is detected. This means a 1:0.005 B$_4$C:Pd^{2+} ratio offers the best Ni-B nanolayer surface coverage of B$_4$C particles. At the other three B$_4$C:Pd^{2+} ratios, the B-C bond can be easily detected, likely due to the uneven Ni-B nanolayer on the B$_4$C particle surfaces, with some locations being thinner and some locations being thicker than the X-ray penetration depth. This observation confirms that the B$_4$C:Pd^{2+} ratio of 1:0.005 is the optimal PdCl$_2$ content for the B$_4$C particle surface activation.

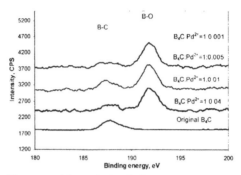

Figure 4. XPS spectra of the original and Ni-B nanolayer coated B$_4$C particles activated using different PdCl$_2$ amounts.

3.2. Complexing Agent Effect

As discussed in our prior work, the reducing agent NaBH$_4$ will only be effectively functioning under a highly alkaline condition.[8] However, when NiSO$_4$·6H$_2$O is dissolved directly into highly alkaline aqueous solution, Ni^{2+} will be released from its water complexed state and form nickel hydroxide sediment, which would make it impossible for the electroless coating to happen. To avoid nickel hydroxide formation, Ni^{2+} must be complexed with a strong complexing agent before the system is adjusted to the high pH condition. According to the complexing mechanism, a 1:3 theoretical molar ratio of Ni^{2+}:C$_2$H$_8$N$_2$ is required to obtain a stable nickel complex. However, our experiments have shown that the Ni^{2-}:C$_2$H$_8$N$_2$ ratio of 1:3 cannot maintain stable complexed Ni^{2+} ions. More C$_2$H$_8$N$_2$ is needed than shown in the above ratio. In this work, the Ni^{2-}:C$_2$H$_8$N$_2$ molar ratios of 1:4.5, 1:6, and 1:9 are studied in order to evaluate the effect of the complexing agent C$_2$H$_8$N$_2$ on the nanolayer coating. The B$_4$C:Pd^{2+} ratio is 0.04 and the NaBH$_4$ addition rate is 10 drops/min.

Figure 5 presents SEM images of the original and coated B$_4$C particles using different amounts of complexing agent. Comparing the original and the coated surfaces, when the Ni^{2+}:C$_2$H$_8$N$_2$ ratio is 1:4.5, a mesh structured layer covers the B$_4$C particle surfaces and a high number of Ni-B nodules can be observed. When the Ni^{2+}:C$_2$H$_8$N$_2$ ratio is 1:6, the mesh structured layer covers the B$_4$C particle surfaces with smaller sized Ni-B nodules. When the Ni^{2+}:C$_2$H$_8$N$_2$ ratio is 1:9, the B$_4$C particle surface is smooth and the sharp edges of the B$_4$C particles can be observed. The particle surface is very similar to the bare B$_4$C surface. and the mesh structure is not visible anymore. This observation implies that increased Ni^{2+} stability and gradual Ni^{2+} release due to stronger complexing from C$_2$H$_8$N$_2$ is beneficial for more uniform Ni-B nanolayer growth. However. when Ni^{2+} is complexed too strongly with the C$_2$H$_8$N$_2$ (1:9 ratio). the release of Ni^{2+} ions into the electroless coating bath can be substantially hindered and the Ni-B nanolayer formation rate becomes very slow. even though the chemical ratios of the electroless coating reaction remain the same.

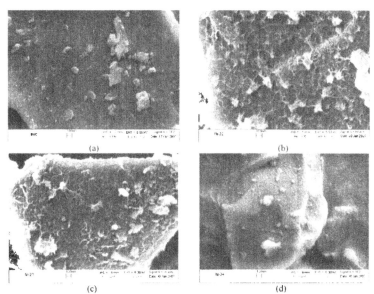

Figure 5. SEM images of the original (a) and coated B$_4$C particles
using different Ni:C$_2$H$_8$N$_2$ ratio: (b) 1:4.5, (c) 1:6, and (d) 1:9.

In order to further examine the surface chemistry. XPS was carried out. To achieve a better visual effect the spectra are separated on the plot. As shown in the right plot of Figure 6. when the Ni^{2+}:C$_2$H$_8$N$_2$ ratio is 1:4.5 and 1:6, both B-C bond and B-O bond are detected. This means the Ni-B nanolayer is present but non-uniform on the B$_4$C particle surfaces. Some areas have less than a 5 nm thickness (XPS penetration limit) and can be penetrated by the X-ray during the XPS analysis. A closer look reveals that when the Ni^{2+}:C$_2$H$_8$N$_2$ ratio is 1:6. the B-C peak is very low with respect to the B-O peak. especially when compared to a 1:4.5 ratio in which case the intensity of the B-C peak is stronger than that of the B-O peak. This means the Ni-B nanolayer coverage on the B$_4$C particles is much improved with the 1:6 ratio. When the Ni^{2+}:C$_2$H$_8$N$_2$ ratio is 1:9. however. the strong B-C peak re-appears. and it is the only peak presenting in the spectra. similar to that of pure B$_4$C. as seen in the left plot of Figure 6. This means the Ni-B coating is either non-uniform or too thin and the XPS analysis detects the B-C bonds from the B$_4$C particles. Combined with Figure 5. it can be concluded that using less amounts of complexing agent facilitates the formation of the Ni-B coating on the B$_4$C particle surfaces. Using excess complexing agent lowers the Ni^{2+} release rate to the electroless coating bath and results in very thin Ni-B nanolayer formation. Even though it is difficult to form stable and complexed Ni^{2+} with too little complexing agent. too much complexing agent is also undesirable since it adversely affects Ni^{2+} availability for the Ni-B nanolayer formation. A balance needs to be achieved between the Ni-B nanolayer morphology and thickness. According to the above discussions, the Ni^{2+}:C$_2$H$_8$N$_2$ ratio of 1:6 is preferred when both SEM and XPS results are considered.

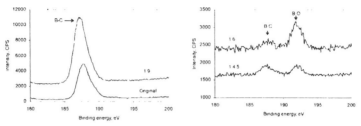

Figure 6. XPS spectra the original and coated B$_4$C particles using different Ni:C$_2$H$_8$N$_2$ ratios.

3.3. Reducing Agent Addition Rate Effect

As mentioned at the beginning, the reducing agent was added finally instead of being mixed with all other components composing the electroless plating solution before the to-be-coated substrate is introduced. As a result, the reduction takes place in a way that the reducing rate is strictly controlled by how the reducing agent is introduced into the system. Since the electroless coating is in fact a chemical reduction procedure, the reducing agent plays a crucial role in the final step. The reducing reaction will compete with the complexing effect from C$_2$H$_8$N$_2$. As discussed above, the amount of complexing agent affects the coating quality. In other words, the nanolayer formation is strongly dependent on the rate that Ni^{2+} can be reduced.

Figure 7 displays SEM images of the bare and Ni-B coated B$_4$C particles with different NaBH$_4$ addition rates. Different NaBH$_4$ addition rates used are 1 drop/min, 10 drops/min, and all at one time. The B$_4$C:Pd^{2-} ratio is 1:0.04 and the Ni:C$_2$H$_8$N$_2$ ratio is 1:6. As shown in Figure 7 (b), when all the NaBH$_4$ is added into the electroless coating bath at one time, the Ni-B layer forms at a fast rate and acts as a 'glue' to agglomerate the B$_4$C particles. Due to the high rate of the reduction and the formation of Ni-B, multiple B$_4$C particles are encapsulated into a flaky cluster but some of the B$_4$C particles inside the cluster are not coated with the Ni-B nanolayer. Figure 7 (c) demonstrates that when the NaBH$_4$ is added with a rate of 10 drops/min, a better defined Ni-B layer forms on the B$_4$C particle surfaces and the glue-like morphology disappears. Finally, when the NaBH$_4$ is added with a rate of 1 drop/min, a thin and uniform Ni-B layer can be observed, as seen in Figure 7 (d).

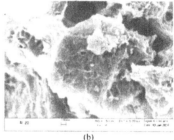

(a)   (b)

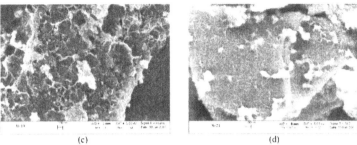

(c)   (d)

Figure 7. SEM images of (a) the original and coated B$_4$C particles with different NaBH$_4$ addition rate: (b) all at one time, (c) 10 drops/min, and (d) 1 drop/min.

XPS spectra of samples obtained when different NaBH$_4$ addition rates are presented in Figure 8. When all the NaBH$_4$ is added at one time, the B-O peak dominates the XPS spectra with a small B-C peak. This implies the B$_4$C particle surfaces are mostly covered with the Ni-B nanolayer, but some locations might have a very thin Ni-B layer, resulting in the X-ray penetrating this thin layer and the original B$_4$C surface being detected. Also, the B-C peak might come from the uncoated but exposed B$_4$C surfaces. When NaBH$_4$ is added at 10 drops/min, the intensity of the B-O peak increases and the intensity of the B-C peak is low. This means a slower NaBH$_4$ addition rate is beneficial for improved Ni-B nanolayer surface coverage. More Ni and B are deposited on the B$_4$C particle surfaces. When NaBH$_4$ is added at 1 drop/min, the B-O peak is still present but the B-C peak dominates the XPS spectra. This is likely because the Ni-B nanolayer is too thin and X-ray penetrates through the thin layer easily, as reflected by the smooth surface shown in Figure 7 (d).

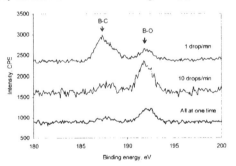

Figure 8. XPS spectra of the Ni-B coated samples when NaBH$_4$ is added with different rate.

Figure 9 shows the atomic ratio between boron in the B-O bond and nickel in the Ni-B nanolayer when NaBH$_4$ is added at different rates. The reason that only boron in the B-O bond is considered is that it has been confirmed by our XPS data that the B-O bond is detected only on the coated sample surfaces. Hence, the B-C bond detected is contributed by the original B$_4$C particle. The B:Ni atomic ratio decreases with the decreasing NaBH$_4$ addition rate. The explanation is as follows. NaBH$_4$ is not only acting as the reducing agent, it will also decompose at the same time. If it is added in a very small amount that is much less compared to the amount of Ni^{2+} ions in the system, it will readily react with Ni^{2+}. However, if NaBH$_4$ is added at a fast rate, some molecules will decompose.

Since we are using more NaBH$_4$ than necessary to completely reduce Ni^{2+} and achieve a coating thickness as close to the targeted value as possible, it is not hard to understand that there is more boron in the case where NaBH$_4$ is added at a faster rate.

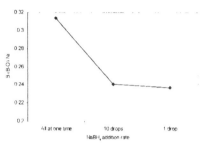

Figure 9. Chemical composition of Ni-B nanolayer when NaBH$_4$ is added with different rate.

After considering the effects of the activation agent PdCl$_2$, the complexing agent C$_2$H$_8$N$_2$, and the reducing agent NaBH$_4$ addition rate, electroless coating was carried out combining the optimal values of these factors. SEM images of the resulting B$_4$C surface and the cross section are shown in Figure 10. A uniform and well defined Ni-B nanolayer is obtained with all three conditions optimized. The mesh-like nanostructures among the uniform and small Ni-B nodules can be seen in the left image of Figure 10. In the right image of Figure 10 it can be seen that the bright Ni-B nanolayers are distributed all around the B$_4$C particle surfaces. The coating looks uniform and integrated. This means Ni-B coverage on the B$_4$C particles is complete.

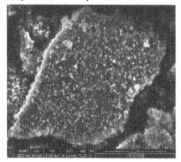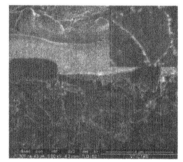

Figure 10. SEM images of the coated B$_4$C surface and cross section
obtained using optimized coating condition.

3.4. Combustion Driven Compaction

Combustion driven compaction (CDC) is a unique technique developed by Utron, Inc. for forming particulate components.[10][11] Different from all other compaction techniques, combustion driven compaction presses powders through the use of high pressure generated by combusting a gas mixture with a simple set up not requiring a large area. A schematic of CDC set-up is shown in Figure 11 (left) and the schematic of the compaction chamber is shown too (Figure 11, right). During the compaction cycle, the chamber is filled with a mixture of natural gas and air. As the chamber is being filled the ram is able to move down, pre-compressing and removing entrapped air from the powder. When the sealed gas mixture is ignited, the pressure in the chamber will rise dramatically in a very short period of time which is normally in the order of milliseconds. At the same time, the ram will be pushed down to the powder at an extremely high speed, realizing the compaction.

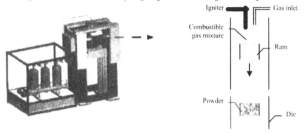

Figure 11. Combustion driven compaction setup (left) and the compaction chamber schematic (right).

Figure 12 shows a typical CDC load cycle and the corresponding microstructure evolution of the compressed powder. During this cycle, a high pressure >3 GPa can be achieved during a very short period of time (less than 200 ms). Also, from Figure 12, it is apparent that the pressure release is relatively slow compared to the compaction (>750 ms *v.s.* <200 ms), which will effectively reduce the localized stress and the resulting defects.

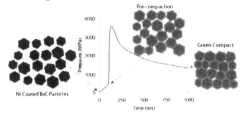

Figure 12. Compaction process of Ni-B nanolayer coated B_4C.

In this work, B_4C/Ni-B particles have been compacted with 2.63 GPa peak pressure. The Ni nanolayer is predicted to act as a lubricant to reduce interparticle frictional force and facilitate B_4C particle sliding and densification. The combination of >2 GPa compaction pressure and the lubricating Ni layer produces a unique opportunity for highly effective B_4C/Ni particle densification. After the combustion driven compaction, it has been observed that a continuous Ni-B network forms in the B_4C compact. The uniform distribution of the Ni-B nanolayer holds the B_4C particles together strongly and prevents defect formation while allowing stress release, which makes it especially attractive when making parts that have to endure many handling processes before being sintered.

Figure 13 shows pictures of compacts obtained from combustion driven compaction. Equibiaxial flexural strength test was carried out to evaluate the compacts. The average green strength is 6.87 MPa. This is very high compared to the conventional B_4C compacts, which can break fairly easily. Also, the green density of the Ni-B nanolayer coated B_4C particles is higher, at ~75% density, in comparison to 68% density achieved under conventional compaction conditions.[12]

Figure 13. Compacts obtained from combustion driven compaction of
Ni-B nanolayer coated B$_4$C particles.

4. CONCLUSIONS

The effects of the activation agent (PdCl$_2$) amount, the complexing agent (C$_2$H$_8$N$_2$) amount, and the reducing agent (NaBH$_4$) addition rates on the electroless Ni coating of micron-sized B$_4$C particles are studied. An optimal condition of each factor is determined utilizing SEM and XPS techniques. The complexing agent C$_2$H$_8$N$_2$ improves the Ni-B nanolayer morphology but lowers its growth rate. The reducing agent NaBH$_4$ addition rate affects the nanolayer thickness and composition. When the optimal combination is used in the electroless coating, the resulting B$_4$C particle is surrounded by a uniform and continuous mesh-structured layer. XPS also reveals that boron is incorporated into the coating in the form of boron oxide. Using the combustion driven compaction technique, compacts with ~75% density are obtained from Ni-B nanolayer coated B$_4$C particles. The green samples show high equibiaxial flexural strength.

5. ACKNOWLEDGEMENTS

This work is supported by National Science Foundation under grant # DMI-0620621.

*Targeted Ni-B nanolayer thickness was calculated based on the NiSO$_4$·6H$_2$O: B$_4$C ratio. Actual thickness is somehow larger than the stated value due to the incorporation of boron and oxidation of nickel and boron as will be discussed later. As figured out in our prior work, the 5 nm targeted thickness offers the best coating morphology and uniformity with the least amount of nickel possible being introduced into B$_4$C. Therefore the amount of NiSO$_4$ corresponding to 5 nm targeted thickness was chosen in this study.

REFERENCES

[1]. J.W. Jung and S.H. Kang, Advances in Manufacturing Boron Carbide-Aluminum Composites, *J. Am. Ceram. Soc.*, **87** (1), 47-54 (2004).

[2] V.V. Skorokhod and V.D. Krstic, Processing, Microstructure, And Mechanical Properties of B$_4$C-TiB$_2$ Particulate Sintered Composites I. Pressureless Sintering and Microstructure Evolution. *Powder Metall. Met. Ceram.*, **39** (7-8), 414-423 (2000).

[3] V.V. Skorokhod and V.D. Krstic, Processing, Microstructure, And Mechanical Properties of B$_4$C-TiB$_2$ Particulate Sintered Composites. II. Fracture and Mechanical Properties, *Metall. Met. Ceram.*, **39** (9-10), 504-513 (2000).

[4] L. Levin, N. Frage and M.P. Dariel, The Effect of Ti and TiO$_2$ Additions on the Pressureless Sintering of B$_4$C, *Metall. Mater. Trans.*, A **30** (12), 3201-3210 (1999).

[5] S. Arai, M. Endo, S. Hashizume and Y. Shimojima, Nickel-Coated Carbon Nanofibers Prepared by Electroless Deposition, *Electrochem. Commun.,* **6** (10), 1029-1031 (2004).

[6] Y.J. Chen, M.S. Cao, Q. Xu and J. Zhu, Electroless Nickel Plating on Silicon Carbide Nanoparticles, *Surf. Coat. Technol.,* **172** (1), 90-94 (2003).

[7] N. Dadvand, G.J. Kipouros and W.F. Caley, Electroless Nickel Boron Coating on AA6061, *Canadian Metallurgical Quarterly* 42 (3), 349-363, 2003.

[8] X. Zhu, H. Dong, K. Lu, Coating Different Thickness Nickel-Boron Nanolayers onto Boron Carbide Particles, *Surf. Coat. Technol.,* 202, 2927–2934, 2008.

[9] S.L. Brandow, W.J. Dressick, C.R.K. Marrian, G.M. Chow and J.M. Calvert, The Morphology of Electroless Ni Deposition on a Colloidal Pd (II) Catalyst, *J. Electrochem. Soc.,* **142** (7), 2233-2243 (1995).

[10] K. Nagarathnam, D. Trostle, D. Kruczynski and D. Massey, Materials Behavior and Manufacturing Aspects of High Pressure Combustion Driven Compaction P/M Components, in: *International Conference on Powder Metallurgy & Particulate Materials*, pp. 1-15, Chicago, IL, June 13-17, (2004).

[11] K. Nagarathnam, A. Renner, D. Trostle, D. Kruczynski and D. Massey, Development of 1000-ton Combustion Driven Compaction Press for Materials Development and Processing, in: MPIF/APMI *International Conference on Powder Metallurgy & Particulate Materials*, Denver, Colorado, (2007).

[12] H. Lee and R.F. Speyer, Pressureless sintering of boron carbide, *J. Am. Ceram. Soc.,* **86** (9), 1468-1473 (2003).

EFFECT OF CARBON NANOTUBES ADDITION ON MATRIX MICROSTRUCTURE AND THERMAL CONDUCTIVITY OF PITCH BASED CARBON - CARBON COMPOSITES

Lalit Mohan Manocha[1,4]*, Rajesh Pande[1], Harshad Patel[1], S. Manocha[1], Ajit Roy[2] and J. P. Singh[3]
[1]Department of Materials Science, Sardar Patel University, Vallabh Vidyanagar 388 120, Gujarat, India
[2]Air Force Research Laboratory, USA
[3] AOARD, Tokyo, Japan
[4] Sophisticated Instrumentation Centre for Advanced research and Testing,
Vallabh Vidyanagar -388 120, Gujarat, India
*E-mail: manocha52@rediffmail.com

ABSTRACT

Carbon/carbon composites were developed with carbon nanotubes as reinforcements and mesophase pitch as matrix precursor. The composites were carbonized at 1000°C and graphitized at 2400°C. Microstructure of these composites has been studied after each heat treatment using XRD, Scanning Electron Microscope (SEM), Raman Microscope and polarized light optical microscope. The composites were also studied for thermal conductivity measurement. The microstructure of pitch based carbon matrix was found to change from anisotropic to isotropic structure with addition of carbon nanotubes but the thermal conductivity of the reinforced composites is not found to change much mainly due to the presence of high thermal conductivity of nanotubes.

1. INTRODUCTION

Carbon nanotubes filled carbon matrix composites with unique combination of physical, mechanical and electrical properties of carbon nanotubes and carbon matrix are expected to be best candidate material for thermal management application[1,2]. This is required a lot of basic studies to be performed on the properties of carbon nanotubes, their bonding with the matrix systems and translation of nanotubes properties to the composites. However, very little is published on effect of nano fillers on microstructure of mesophase pitch based carbon/carbon composites and their composites[3-8]. The objective of this research work was to study dispersion of multiwall carbon nanotubes in the pitch based matrix precursors. The microstructure of pitch based carbon matrix and their influence on the thrmal conductivity of ultimate carbon-carbon composites.

2. EXPERIMENTAL

2.1. Materials Used

Multiwall carbon nanotubes of 20-60nm diameter and 85- 100 micron in length were used as reinforcement material. These nanotubes were synthesized on flat quartz substrate by decomposition of ferrocene-xylene solution at 800 °C. Which were then purified by 2.5 % V/V nitric acid solution. Fig.1. shows the SEM micrograph of purified carbon nanotubes used in present studies.

Matrix precursors used for making composites was AR mesophase pitch (Mitsubishi Gas Chemical Company Inc. Japan). The characteristics of AR mesophase pitch used for making these composites are compiled in Table.I.

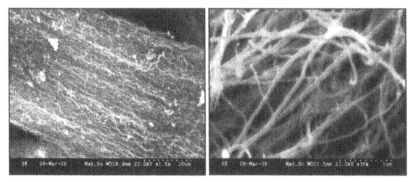

Fig. 1. SEM micrograph of acid purified carbon nanotubes

Table -I Characteristics of matrix precursor

	AR Mesophase pitch
Softening point (°C)	275
Coke yield (%)	63
QI content (%)	20.57

2.2. Processing of Carbon/Carbon Composites

The carbon nanotubes were closely packed within the bundles. In order to achieve better dispersion of these nanotubes in the pitch matrix, the carbon nanotubes films were ultrasonicated in acetone following by drying at 70 °C. 1 wt % CNTs were mixed with the fine pitch powder and homogenized in ball mill. This ball milled powder mixture was palletized in hot press at 280 °C. Pallet of as such mesophase pitch was also prepared in the same way. The green samples were stabilized at 200 °C for 24 hrs in air atmosphere at normal pressure. These composites were carbonized in two-steps. The green pallets were first carbonized to 550 °C at 10 °C/hour and then further heated to 1000 °C under inert atmosphere of nitrogen. These composites were further heat treated at 2400 °C under normal pressure of argon.

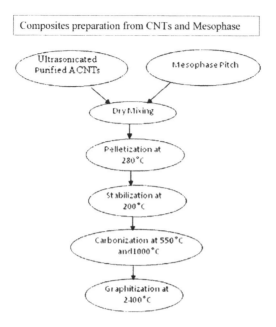

2.3 Studies on microstructure of the composites

Microstructure of the carbon matrix in composites was studied by viewing the highly polished composites under polarized light using Leitz Optical Microscope 12 Pol S. These composites were also viewed using Hitachi S 3000 N scanning electron microscope to studied dispersion and bonding of carbon nanotubes with carbon matrix.

Raman spectrometer (Renishaw InVia Raman microscope) was used to evaluate the structure of the pitch alone heat treated to different temperature as well as the composites containing carbon nanotubes. These spectra were recorded using argon gas laser (514 nm excitation wavelength), which corresponds to the equivalent photon energy of 1.58eV. The x-ray diffractograms were obtained by X-ray diffraction ((Phillips X`pert) using CuKα radiation (1.54056 Å) to determine crystalline parameters (crystallite size, d-spacing and degree of graphitization) of nanocomposites.

Thermal diffusivity of the heat treated pitches as well as the carbon-carbon composites with carbon nanotubes as reinforcement was measured at room temperature by laser flash method on NEITZSCH Micro flash- 300. This technique employs a high-power infrared pulse laser to deliver a short burst of energy to one face of a disk shaped specimen. The resulting rise in temperature on the opposite face is monitored by an infrared detector and recorded as a function of time by a computerized data acquisition system. The thermal conductivity, k, can then be calculated by

$$k = \alpha\, C_p\, \rho$$

Where α is the thermal diffusivity, C_p is the specific heat and ρ is the density of the composite.

3. RESULTS AND DISCUSSIONS

3.1. Microstructure of Carbon Nanotubes reinforced Carbon composites

Fig. 2 shows the optical micrograph of the as such pitch derived carbon as well as of their composites with carbon nanotubes. The presence of nanotubes found to affects the texture of the carbon matrix. These seem to be aroused from the lack of long-range shearing of mesophase due to presence of nano sized filler. Also the nanotubes are well-distributed and randomly oriented in the matrix. These nanotubes will suppress the mesophase formation in pitch resulting in isotropic mosaic structure of the ultimate carbon matrix. This phenomena was also supported by the earlier work of the group[7] in which they showed that the milled fibers having longer length do not disturb the matrix texture but the small

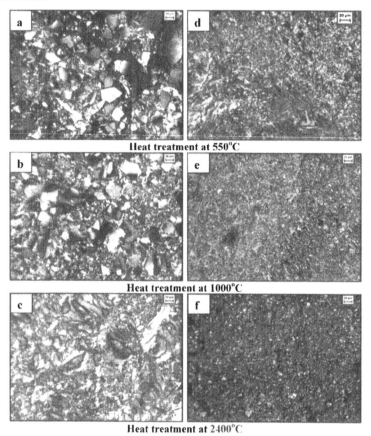

Heat treatment at 550°C

Heat treatment at 1000°C

Heat treatment at 2400°C

Fig. 2. Optical photographs of as such mesophase derived carbon (a. b and c) and 1 wt% CNTs added carbon matrix (d. e and f) after 550. 1000 and 2400°C heat treatment.

irregular sized milled fibers and dusts of milled fibers (size: ⁻2 µm) disturb orientation of the matrix structure.

The SEM micrographs of these composites show good bonding between nanotubes and matrix after carbonization. The same is found to be retained even after heat treatment of 2400 °C. As such mesophase pitch based carbon shows layered microstructure after 2400 °C heat treatment. The nanotubes added composites exhibit good bonding between nanotubes and carbon matrix and CNTs seem to be well distributed. The fracture surface of 2400°C heat-treated composite shows good bonding of CNTs with carbon matrix.

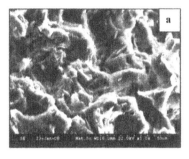

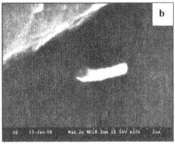

Fig.3. SEM micrographs of (a) as such mesophase pitch based carbon heated at 2400 °C and (b) mesophase pitch-nanotubes composite heat treated at 2400°C.

The transformation in matrix microstructure is also evident from the x-ray analysis. Fig.4 shows XRD patterns of the carbon matrix and of the carbon-carbon composites. Table - 2 shows the results of XRD analysis of carbon matrix alone as well as of the composites. The crystallinity of matrix materials is observed to be affected by the nano reinforcements. The value of d- spacing for composite (0.3408 nm) is found to be higher than that of as such mesophase based carbon (0.3385 nm). The degree of graphitization was found to decrease from 63.95 to 32.20.

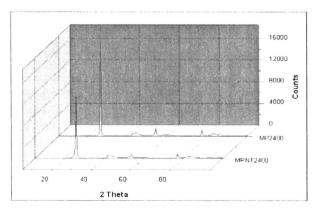

Fig.4. XRD spectra of mesophase based carbon (MP2400) and
nanocomposites (MPNT2400) after 2400°C heat treatment.

Fig.5 shows the Raman spectra of pitch based carbon and of carbon-carbon composites. Typical
peaks of disordered (D band) and ordered (G band) carbons were observed for these samples. These
Raman lines are attributed to lattice defects and finite crystal size of carbon microstructure[9]. Fig.5 (a)
shows the Raman

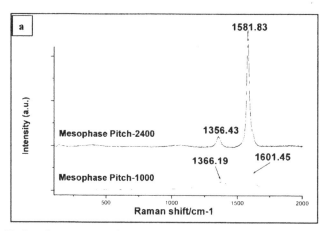

Fig.5 (a) Raman spectra of mesophase pitch based carbon after Carbonization
and Graphitization.

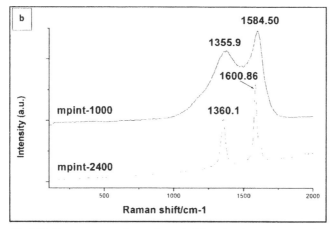

Fig.5 (b) Raman spectra of mesophase pitch based carbon nanocomposites after Carbonization and Graphitization.

lines at 1366.19cm^{-1} and 1601.45 cm^{-1} for mesophase derived carbon after carbonization. The ratio of I_D/I_G for this sample was 0.41. This ratio decreases after graphitization to 0.09 for 1356.43 cm^{-1} and 1581.83 cm^{-1} for spectral lines, which shows better crystallinity.

In case of composites, the Raman lines were observed at 1355.9 cm^{-1} and 1584.5 cm^{-1} for carbonized samples and at 1360.1 cm^{-1} and 1600.86 cm^{-1} for graphitized samples. The I_D/I_G ratio for graphitized sample was 0.61. Composites shows high I_D/I_G ratio this also supports the addition of nanotubes into pitch suppresses the graphitization process.

3.2. Thermal Conductivity of Carbon/Carbon Composites

Room temperature thermal conductivity of composites made under present studies has been compiled in Table II. In case of pitch as such, the conductivity was measured in the direction of anisotropy (graphitic orientation) whereas in case of composites, the conductivity was measured in thickness direction of the composites. The carbon nanotubes are randomly distributed with most of the tubes having long side perpendicular to the thickness.

Table -II Properties of carbon Nanocomposite

Sample	d-spacing A^0	Degree of graphitization gp %	Thermal conductivity W/m. K.
Mesophase pitch HTT 2400°C	3.38548	63.95	43.82
Mesophase pitch + 1 wt % CNTs HTT 2400°C	3.40852	32.20	40.38

Thermal conductivity of mesophase pitch based carbons was found to be 43.82 W/m. K. and thermal conductivity of nanotubes added composite was found to be 40.38 W/m. K. The small decrease in thermal conductivity of nanocomposites may be due to the disturbance of orientation of graphitic layers by the presence of nanotubes. Though the microstructure of carbon matrix was found to change from anisotropic to more isotropic in nature, the change in thermal conductivity is found to be marginal.

4. CONCLUSIONS

Carbon Nanotubes as reinforcement are very efficient in controlling the microstructure of pitch based carbon matrix and properties of the composites. 1 wt % addition of nanotubes in mesophase pitch disturbed alignment of graphitic structure of the matrix and decreased the degree of graphitization of composites. But the thermal conductivity was found to remain almost same.

ACKNOWLEDGEMENTS

The authors are thankful to The Department of Science and Technology, Government of India for its financial support for this research work under research project No. SR/S5/NM-52/2003 as well as AOARD, Contract No: 05-4092. The authors are also thankful to Sophisticated Instrumentation Centre for Applied Research & Testing (SICART) for X-Ray diffraction and thermal conductivity measurement.

REFERENCES

[1]E. T. Thostensona, Z. Renb, and T. W. Choua, "Advances in the Science and Technology of Carbon Nanotubes and their Composites: a review", Composites Science and Technology, 61 (2001) 1899–1912

[2]M. Terrones, "Science and Technology of the Twenty-First Century: Synthesis, Properties and Applications of Carbon Nanotubes, Annu. Rev. Mater.Res. 33 (2003), 419-501

[3]T. Cho, Y. Lee, S. R. Rao, A. M. Rao, D. D. Edie, and A. A. Ogale, "Structure of Carbon Fiber Obtained from Nanotube-Reinforced Mesophase Pitch", Carbon, 41/7, (2003) 1419-24

[4]R. Andrews, D. Jacques, A. M. Rao, T. Rantell, F. Derbyshire, Y. Chen, J. Chen and R. C. Haddon, "Nanotube Composite Carbon Fibers", Appl. Phys. Lett., 75, (1999), 1329

[5]I. Pasuk, C. Banciu, A.M. Bondar, G.A. Rimbu, I. Ion, I. Stamatin, I. Morjan, I. Voicu and I. Sandu; "Influence of some Carbon Nanosrustures on the Mesophase Pitch development- A Structural Study " 6Romanian Reports in physics 56/3. (2004), 320-327

[7] L. M. Manocha; A. Warrier, S. Manocha, D. Sathiyamoorthy and S. Benerjee, "Thermophysical properties of densified pitch based carbon/carbon materials—II. Bidirectional composites" Carbon, 44 (2006) 488-495

[8]L.M. Manocha, A. Warrier, S. Manocha, D. Sathiyamoorthy and S. Benerjee, "Thermophysical Properties of Densified Pitch based Carbon/Carbon materials- I. Unidirecttional Composites" Carbon, 44 (2006) 480-487

[9]M.S. Dresselhaus, M.A. Pimenta, P.C. Eklund and G. Dresselhaus, "Raman Scattering in Fullerenes and related Carbon-based Materials. Raman Scattering in Materials Science. Springer Series in Materials Science, vol. 42, WeberWH, Merlin R (eds). Springer.

MICROSTRUCTURE AND PROPERTIES OF CARBON NANOTUBES REINFORCED TITANIA MATRIX COMPOSITES PREPARED UNDER DIFFERENT SINTERING CONDITIONS

S.Manocha[1] , L.M.Manocha*[1], E.Yasuda[2], Chhavi Manocha[3]

[1] Department of Materials Science, Sardar Patel University, Vallabh Vidyanagar -388120, India

[2] Materials and Structures Laboratory, Tokyo Institute of Technology, Yokohama, Japan

[3] GCET Engineering College, Vallabh Vidyanagar, India

*E-mail: manocha52@rediffmail.com

ABSTRACT

The properties of titanium based ceramics can be tailored either through control of microstructure at nanoscale or through addition of nano reinforcements to have prospective material for structural applications and as biomaterial. Amongst nanoscale reinforcements, carbon nanotubes, being the most promising materials as reinforcements for ceramics, present studies were undertaken to develop Carbon nanotubes reinforced Titania matrix composites and to study its microstructure and properties.

Multiwall carbon nanotubes (CNTs) synthesized by CCVD technique were used as reinforcements. Titania matrix was prepared using sol-gel technique. CNTs were coated with Titania sol using spray drier. CNTs Titania ratio was kept as 1:100. Composites of titania/carbon nanotubes were prepared through sintering under normal pressure as well as under hot press at $1400^{\circ}C$ under 33 KPa pressure. Sol-gel technique coupled with spray dryer resulted in uniform distribution of CNTs in Titania matrix. Titania ceramic composites prepared using carbon nanotubes as reinforcements result in change in physical properties, enhanced compressive strength. These results show that Hot Press Sintering technique results in composites with reduced porosity and about 20% enhancement in compressive strength.

1. INTRODUCTION

Titanium metal and its alloys, possessing excellent structural properties and chemical corrosion resistance, are used mainly by the aerospace and chemical industries. It is also used in ceramic form either as carbides or oxides. Though maximum of titania is used in paints industries and cosmetics, for electrical uses in electronic industries, specifically as sensors, as photosensitiser for photovoltaic cells and as electrocatalyst [1,2], it is also prospective material for structural applications and as biomaterial. The properties of these ceramics can be improved either through control of microstructure at nanoscale as well as through addition of reinforcements. Toughness of such micromaterial reinforced ceramics could be increased but at the cost of strength both at room temperature and at elevated temperatures [3]. Moreover, nanostructured materials have attracted the structural scientists, engineers and industries with simple promise that using building blocks with dimensions in the nanosize range makes it possible to design and develop new multifunctional materials, the so called nanocomposite materials with unique combination of properties unachievable with traditional materials. Amongst nanoscale reinforcements, carbon nanotubes having exceptionally high mechanical properties and high temperature stability are the most promising materials as reinforcements for ceramics (4, 5). Proper dispersion of nanomaterials into the matrix is of major concern. This can be better achieved through solution route of processing of ceramics. Therefore, present studies were undertaken to develop Carbon nanotubes reinforced Titania matrix composites using sol-gel technique and to study effect of various processing parameters on sintering and microstructure and properties of the ultimate composites.

2. EXPERIMENTAL

Following were the materials used in the present studies:

\# Titanium Isopropoxide as the precursor (Fluka make)

\# Laboratory grown Multi walled Carbon Nanotubes as reinforcement

2.1. Carbon Nanotubes

Multi wall carbon nanotubes used in the present studies were synthesized in Department of Materials Science, Sardar Patel University, using CCVD technique with Xylene as carbon precursor and Ferrocene as catalyst precursor (5). Fig.1 shows the SEM photographs of the carbon nanotubes. The diameter of these nanotubes was 50-80 nm and these were 2-5 microns long.

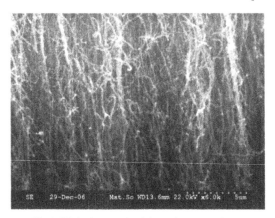

Fig.1 SEM micrograph of the carbon nanotubes

2.2. Preparation of Titania sol

Titania sol was prepared by the hydrolysis of Titanium Isopropoxide. Titanium Isopropoxide and Isopropanol were taken in the stoichiometry ratio of 2:1 in a beaker and were mixed with the help of a stirrer. When the solution became clear, hydrolysis was carried out with distilled water at pH 2. Titanium hydroxide formed due to hydrolysis of titanium isopropanol. The mixing was carried out for four hours.

2.3. Coating of Carbon Nanotubes with Titania sol.

Carbon nanotubes were uniformly coated with Titania sol using Spray Drier. The amount of sol as well as the processing conditions in the spray drier was so adjusted to have ultimately 1% (by weight) of carbon nanotubes in the titania sol. These were post dried at $100^{\circ}C$. The samples collected from the drier were in the form of very fine powder.

2.4. Pressureless Sintering of Titania – Carbon Nanotube Composites

The Titania sol-carbon nanotubes powder was compressed in a die under pressure of 17 kg/cm^2. These composites, called green composites were obtained in the form of round pallets. These composites were sintered at $1400^{\circ}C$ under nitrogen atmosphere. The heating rate from room temperature to $1400^{\circ}C$ was kept at $100^{\circ}C/hr$. The composites obtained were compacted solid materials. These composites are designated as Composites A.

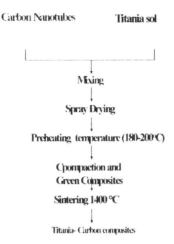

Fig.2. Schematic presentation of the scheme for preparation of composites.

2.5. Hot Press Sintering of Titania / Carbon nanotubes
   A part of the green composites were heat treated at 200°C for 24 hours to remove residual moisture. After drying, these were heat treated to 1400°C in Shimadzu Hot Press. The hot press sintering conditions were:
Heating Rate 10° K/min
Temperature attained: 1400° C.
Sintering Pressure: 33 MPa for 12 hrs.
The samples obtained were much more compact and dense as compared to Pressure less sintering.

2.6. Characterization of the Sintered Composites

2.6.1 Determination of Density and Porosity
   Physical density of the composites was calculated by measuring weight of the sample and physical dimensions, area and thickness. Apparent density and open porosity of the samples was measured using Shimadzu Accu Pycnometer 1330.

2.6.2 Scanning Electron Microscopy
   Microstructure of the composites was studied using Hitachi Field Emission Scanning Electron Microscope. FE-SEM gives much better view of the pores as well as of grains.
The composites were examined using FE-SEM in order to study:
Distribution of carbon nanotubes in the ceramic matrix
Bonding between carbon nanotubes and the matrix
Grain structure of the matrix and hence sintering of the matrix as well as of reinforcement and matrix
Voids / Pore structure in the matrix.

2.6.3 XRD studies
   The crystalline phase of the composites was identified by X ray diffraction technique using Phillips X-ray Differactometer.

2.6.4. Compressive strength

Compressive strength of the composites was measured on Instron Universal Testing Machine.

3. RESULTS AND DISCUSSION

3.1. Structural studies

Density of the composites prepared without and with pressure sintering is compiled in table-1.

Table I. Density and Porosity of composites prepared in present study

Sr. No.	Sample No.	Description of sample	Density	porosity
1	TiO2	TiO_2 powder compact Normal sintered	3.652 g/cc	10.2%
2.	A.	TiO_2 + CNTs Normal sintered	3.5987 g/cc	9.4%
3.	B	TiO_2 + CNTs Hot Pressed sintered	3.7375 g/cc	7.8%

Porosity of the samples is also included in this table. As seen from the table, density of hot pressed sintered composites is higher than as such sintered composites. It is obvious because hot pressing reduces the interparticle distance and hence results in better sintering. That is why as such sintered composites though have only about 5% lower density as compared to HP samples, but the open porosity in former composites is about 25% higher than that in HP samples.

3.2. Microstructural Studies

Microstructure of the composites such as distribution of Carbon nanotubes in the titania matrix as well as bonding of the nanotubes with the matrix, deformations in the nanotubes etc. were studied using SEM. Fig. 3 shows SEM micrographs of the two composites prepared under different sintering conditions. As seen from the figures and comparing Fig.3a and Fig. 3c. it is seen that composites made without pressure sintering, though do not show big size voids but exhibit equidistributed macropores. Comparing Figs 3b and Fig. 3d. it is revealed that the grains are closer in pressure sintered composites. These composites exhibit larger grains as compared to those in as such sintered composites.

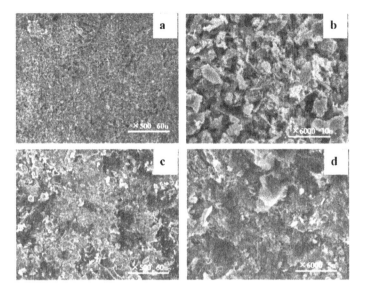

Fig 3. SEM photographs of CNTs/Titania composites.(a and b) Composites
sintered without pressing; and (c and d) composites made through Hot Pressing

3.3. X-ray Diffraction (XRD) Studies

XRD patterns of pure titania show strong peaks at $2\theta = 27.6096$ and $2\theta = 54.2343$ corresponding to Anatase structure and at $2\theta = 41.41$ due to Rutile structure. Fig.4. shows XRD patterns of the two composites. These composites exhibit peaks of carbon at $2\theta = 26.5$ corresponding to carbon nanotubes in addition to the peaks due to pure titania. It is notable to see that composites prepared at normal pressure show peaks due to titanium carbides as well. This means that composites prepared without pressure exhibit reaction between titania and carbon resulting in formation of titanium carbide. However, composites prepared using Hot Pressing technique do not show peaks due to titanium carbide. It means, when sintering is done under pressure, formation of titanium carbide is suppressed.

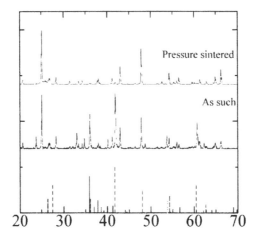

Fig.4. XRD Patterns of As such sintered and Hot Pressed Composites

3.4. Compressive Strength of the Composites

Composites made with 1% addition of carbon nanotubes exhibited about 10-12% higher compressive strength as compared to titania prepared under similar conditions. Composites made using hot pressure technique exhibit about 20% increase in strength. This is quite significant improvement for such small amount of CNT loading. Moreover, the fracture was found to be mixed mode which was an improvement over brittle fracture of ceramics. It shows significance of the addition of CNT to ceramic matrix. The compressive strength shown in table-II

Table II. Compressive strength of the processed composites

Sr. No.	Sample No.	Description of sample	Compressive Strength MPa
1.	TiO_2	Normal sintered	410
2.	A.	TiO_2 + CNTs Normal sintered	450
3.	B	TiO_2 + CNTs Hot Pressed sintered	490

3.5.Vickers Hardness

Fig. 5 shows photograph of the hot pressed composite after Vicker indentation. It also shows non-catastrophic nature, though the composite was not very hard.

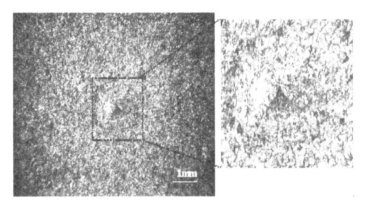

Fig.5. Composites after Vickers Hardness Test at 10 Kgf load

CONCLUSION

Following conclusion could be drawn from the present studies:

Sol-gel process is the most suitable process to control the grain size and hence microstructure of the ceramics and ceramic composites.

Distribution of Carbon nanotubes in ceramic matrix can be better controlled using Spray Dryer Apparatus.

Addition of carbon nanotubes results in enhancement of compressive strength as well as change in fracture behaviour of ceramics.

REFERENCES

[1] K.H.Jurgen Buschow (Editors) "Encyclopedia of Materials: Science and Technology" Elsevier (Publishers) 2001 vol. vol. 7. PP. 6205. vol. 10. PP. 9366-9370

[2] Brinker " Sol gel processing" Academic Press(1986)

[3] W. Krenkel, R.Naslain, High Temperature Ceramic Matrix Composites. Wiley-VCH, (2001)

[4] K.Nihara. (Ed.) *High Temperature Ceramic Matrix Composites III, Trans*-Tech Publication (1998)

[5] L.M.Manocha. J. Valand, S.Manocha and A. Warrier. Transactions of Materials Research Society of Japan.Vol. 29 (8), 3569-3572, 2004.

ELABORATION OF ALUMINA-YAG NANOCOMPOSITES FROM PRESSURELESS
SINTERED Y-DOPED ALUMINA POWDERS

Paola Palmero, Laura Montanaro
Dept. Materials Science and Chemical Engineering, INSTM – R.U. PoliTO - LINCE Lab.,
Politecnico di Torino
Torino, Italy

Claude Esnouf, Gilbert Fantozzi
Université de Lyon, INSA-Lyon, MATEIS CNRS UMR 5510
Villeurbanne, France

ABSTRACT
    The elaboration of 95 vol.% Al_2O_3 – 5 vol.% YAG ($Y_3Al_5O_{12}$) nanocomposites was pursued
by a novel processing route. A commercial nanocrystalline transition alumina powder was doped by
using a yttrium chloride solution. After drying, salt decomposition and final phases development
were obtained by performing a high-temperature treatment, leading to a composite powder made of
α-alumina and YAG. The as-dried material as well as some powders calcined at some selected
temperatures were characterized by means of HR-TEM to follow the morphology and
crystallography evolution as a function of the calcination temperature and time. In addition, the
yttrium distribution was investigated by systematic EDX analyses performed on both bulk and
surface of the alumina particles.
    Differently calcined powders were uniaxially pressed in bars and the influence of the pre-
calcination treatment on the pressureless sintering behaviour was investigated by dilatometric
analyses performed up to 1500°C for 3 hours soaking time. SEM and ESEM observations,
performed on fracture surfaces of sintered bodies, allowed to characterize the final microstructure
of the micro-nano composites, made of equiaxial YAG particles with a mean size of 250 nm,
uniformly dispersed into a submicronic-grained alumina matrix.

INTRODUCTION
    Many studies in the last years have been focused on particulate-dispersed alumina-based
composites, in order to improve both room and high temperature mechanical properties[1-4].
Secondary phases in ceramics have several advantages: they can produce relevant microstructural
modifications (through the suppression of grain growth and the control of grain morphology),
leading to changements in the properties of the monolithic material[5-6]. Enhanced creep behaviour
can be induced by secondary phases giving rise to a strong interface with the matrix[7-13]. Selection of
reinforcing phases having slightly different thermal expansion coefficient than the matrix could
promote thermal stresses at the grain boundaries, modifying the crack propagation path[14].
Moreover, well-dispersed nanosized particles can improve the toughness of the ceramic matrix
through deflection, microcracking, and grain bridging mechanisms[4,15-19].
    Among the various ceramic reinforcing phases, yttrium aluminium garnet (YAG) has been
recently exploited thanks to its excellent creep resistance and high melting point[13].
    Alumina-YAG composites are frequently prepared by mechanical mixing, but this route
does not allow an easy control of the 2nd phase particles distribution in the matrix[20]. For increasing
purity as well as the phase distribution homogeneity, wet-chemical syntheses are also applied, such
as sol-gel[21-22] or co-precipitation of Y and Al ions from a water solution under controlled
conditions[20,23-25]. Recently, the post-doping of an alpha-alumina powder by using yttrium
methoxyethoxide has been successfully exploited to prepare alumina-YAG micro-nanocomposite
materials[11,26]. The precipitation of YAG particles at the grain boundaries and at multiple grain
junctions, of an alumina matrix doped with a yttrium salt, was already observed in literature[27-30].
However, previous papers were mostly devoted to investigate the influence of a low yttrium content

on densification, grain growth[27, 31] and creep behaviour of α-Al_2O_3[7, 29], as a result of yttrium segregation at alumina grain boundaries[29,32,33].

This work discusses the processing and fabrication of 95 vol.% Al_2O_3 – 5 vol.% YAG material, by exploiting an aqueous solution of yttrium salt as a dopant source for a commercial nano-alumina powder.

MATERIALS AND METHODS

Commercial nanocrystalline transition alumina powders (NanoTek by Nanophase Technology, USA, 99.95% purity, average crystallite size of 47 nm[34]) and $YCl_3 6H_2O$ (Aldrich, CAS. N° 10025-94-2, 99.99% purity) were employed for the preparation of 95 vol.% Al_2O_3- 5 vol.% YAG (AY95) powders.

Many aqueous NanoTek alumina slurries (solid content of 33 wt.%) were prepared and powder de-agglomeration was obtained under magnetic stirring for several hours (up to 120 hrs) or by ball milling for 2.5 hrs by using alumina spheres (powder/spheres weight ratio of 1:10). Alpha-alumina spheres having a mean diameter of 2 mm were used. The particle size distribution as a function of the dispersion time was followed by means of a laser particle size analyzer (Fritsch model Analysette 22 Compact). Yttrium doping was performed by drop-wise addition of an aqueous solution of yttrium chloride to the above dispersed suspensions. The doped slurries were then kept under magnetic stirring for 2 additional hrs, before drying in an oven at 105°C or spray drying (Büchi Mini Spray Dryer B-290).

Powders were then submitted to various thermal pre-treatments to induce chlorides decomposition as well as the solid state reaction to yield the final phases, α-Al_2O_3 and YAG. X-Ray diffraction analysis (XRD, Philips PW 1710) was performed on the calcined materials to investigate the phase evolution. The morphological and crystallographic features of differently calcined doped powders was followed by High-Resolution Transmission Electron Microscopy (HR-TEM, Jeol 2010 FEG) observations and elemental analyses were performed by Energy Dispersive X-Ray Spectroscopy (EDX INCA System) with nanoprobes to detect the yttrium distribution on doped samples.

Green bodies were prepared by uniaxially pressing at 300 MPa the pre-treated powders and their sintering behaviour was investigated by dilatometric analysis (Netzsch 402E). Various sintering cycles were tested, by varying the heating rate (from 10°C/min to 1°C/min), as well as the maximum sintering temperature (in the range 1400°-1500°C) whereas a constant soaking time of 3 hrs at the maximum temperature was adopted. The sintering behaviour of the composite samples was compared to that of pure NanoTek, submitted to the same sintering cycles.

The microstructure of fired bodies was then characterised by Scanning Electron Microscopy (SEM , Hitachi S2300) and Environmental Scanning Electron Microscopy (ESEM, FEI XL30 ESEM FEG) observations performed on un-treated fracture surfaces.

RESULTS AND DISCUSSION

NanoTek powder, labelled A, is a mixture of transition aluminas, δ-Al_2O_3 and δ*-Al_2O_3[35], as detected by XRD, and presents some agglomerates having size of about 1.7, 5.5 and 10.4 μm, corresponding to 10, 50 and 90% of the cumulative volume distribution, respectively. The powder was dispersed under magnetic stirring for 120 hrs or by ball milling for 2.5 hrs, in order to reach similar particle size distributions. In Figure 1, the cumulative size distributions by volume of the dispersed samples are compared to that of the as-received material.

NanoTek suspensions were then doped with an yttrium chloride aqueous solution to prepare a 95 vol.% Al_2O_3 – 5 vol.% YAG composite powders (AY95). A doped powder, dispersed under magnetic stirring, was submitted to DSC-TG analysis and its thermal behaviour was compared to that of pure NanoTek, dispersed under the same conditions. The DSC curves recorded in the high temperature regime (700°-1400°C) are shown in Figure 2: the exothermal signals, imputable to α-Al_2O_3 crystallization, are located at about 1190°C and 1275°C for pure and doped samples, respectively. A clear effect of the dopant in increasing the crystallization temperature was therefore observed.

AY95 powder was submitted to various thermal pre-treatments, precisely at 600°C for 0.1 hr, at 900°C, 1150°C and 1300°C for 0.5 hr and finally at 1450°C for 2 hrs, and then characterized by XRD analyses. In all the samples treated at lower temperatures (up to 1150°C), only transition aluminas were detected, probably due to the low percentage of the YAG phase (5 vol.%) in the composite material. Precisely, δ-Al$_2$O$_3$ and δ°-Al$_2$O$_3$ were present in the as-dried, 600°C and 900°C-calcined materials, while θ-Al$_2$O$_3$ phase was also present in the 1150°C treated sample. In this material, α-Al$_2$O$_3$ was not detected, in contrast to pure-NanoTek, which already yielded α-phase traces at the same temperature, thus confirming the role of yttrium in delaying crystallization.

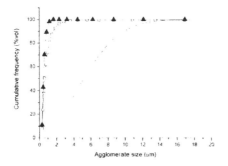

Figure 1. Cumulative size distributions by volume of the as-received NanoTek (solid line without symbols) and of the magnetically stirred (squares) and ball milled (triangles) powder suspensions.

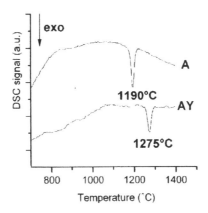

Figure 2. High-temperature DSC curves of pure-NanoTek (A) and of AY95 sample (AY)

Yttrium aluminates and α-Al$_2$O$_3$ were detected in AY95 starting from 1300°C, as shown in Figure 3. At this temperature, YAG and α-Al$_2$O$_3$ were predominant, but also traces of orthorhombic perovskite YAlO$_3$ were observed; on the contrary, after calcination at 1450°C for 2, the material was purely composed by YAG and α–phase.

In order to confirm the above phase development and to investigate the powder morphology evolution as a function of firing, the same thermally treated powders were submitted to HR-TEM observations.

Figure 4 (a) reports the micrograph of the as-dried doped powder, showing the ultra-fine particle size of NanoTek alumina, characterized by an equiaxial shape, as commonly observed for powders prepared by Physical Vapour Synthesis[36].

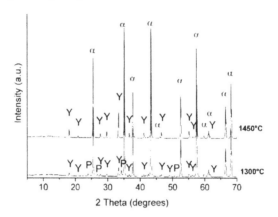

Figure 3. XRD patterns of AY95 powder calcined 1300°C for 0.5 hr and at 1450°C for 2 hrs
($\alpha=\alpha$-Al_2O_3, Y=YAG, P = orthorhombic perovskite $YAlO_3$)

The higher magnification image performed on the same material (Figure 4 b) allows to evidence a thin superficial bad-ordered layer on the first atomic planes, if compared to the well-crystallized inner parts. The related Fast Fourier Transform is reported as an insert in Figure 4 (b) and the spots belonging to the crystalline reflections allowed to identify the δ-Al_2O_3 phase. Alignments of closed-packed spots are attributed to a high density of structural defects inside the transition alumina lattice.

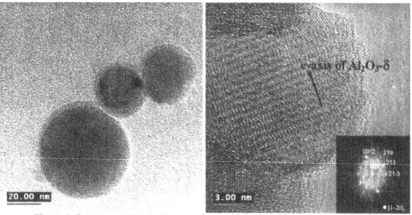

Figure 4. Low (a) and high (b) magnification HR-TEM images of as-dried AY95 powder; the insert of Figure (b) is the indexed Fast Fourier Transform related to the same image.

In addition, the superficial layer was yttrium-rich, as confirmed by systematic EDX analyses performed with a nanoscale probe on both particles boundary and bulk. On the contrary, yttrium was not detected in the bulk of the particles. These microstructural and compositional features were also observed in powders calcined at higher temperature, precisely at 600°C for 0.1 hr and 900°C for 0.5 hr, in which an important increase of yttrium concentration at the surface, in comparison with the particles bulk, was systematically detected.

On the contrary, starting from calcination at 1150°C for 0.5 hr, a little change in the alumina morphology was observed, since well-facetted particles were identified (Figure 5 a), but no crystalline compounds were detected on surface. As clearly shown by the high-magnification micrograph of Figure 5 b, they are again characterized by a bad-ordered surface.

Several EDX comparative analyses were performed by focusing the nanoprobe over the layer and then on an adjacent region within the grain, at few (5-7) nanometres from the previous points. Yttrium was statistically detected on the layer, while its concentration in the inner region was under the instrumental detection limits, thus denoting that even after calcination at 1150°C, yttrium diffusion into the alumina particles was still negligible. The insert of Figure 5 b is the indexed Fast Fourier Transform related to the same image; the atomic planes distances, experimentally determined by the crystalline reflections, allows to identify the δ^*-Al_2O_3 phase. Indeed, the proposed indexation on FFT image corresponds to the best coherent appellation for spots in comparison of all possibilities induced by others phases as γ or δ or θ or ϵ or ζ alumina. Here, δ^*-Al_2O_3 phase will have the following parameters: $a \approx b \approx 0.794$ nm and $c \approx 1.06$ nm (for comparison, according to JCPDS file 46-1215, δ^* lattice parameters are: $a = 0.7934$ nm; $b = 0.7956$ nm; $c \approx 1.1711$ nm).

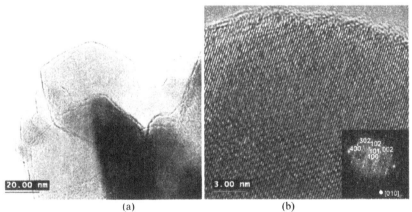

(a) (b)

Figure 5. Low (a) and high (b) magnification HR-TEM images of AY95 calcined at 1150°C for 0.5 hrs; the insert in (b) is the indexed Fast Fourier Transform related to the same image

By calcining at 1300°C for 0.5 hr, a partial sintering of primary particles was yielded, leading to large agglomerates entrapping a diffuse nanoscopic porosity, as shown by the HR-STEM micrograph in Figure 6, performed by using High Angle Annular Dark Field (HAADF) detector.

HAADF imaging is sensitive to the atomic number variations (Z contrast), thus allowing to associate the lighter grains to yttrium-containing phases, whose size was in the range 50-100 nm and which presented an almost equiaxial shape.

In Figure 7, a TEM micrograph of the same material is reported, showing a rounded YAG particle growing on the surface of an alumina grain, as confirmed by EDX analyses (spot size $\approx$ 3 nm) performed on the crystallite (point A in Figure 7a) and on the surrounding matrix (point B). In

Figure 7 b, the EDX peaks intensities of Y and Al determined on points A and B are represented and the atomic Y/Al ratio of 0.6 determined in point A allows to associate the crystallite to the 3:5 molar ratio, garnet-type compound.

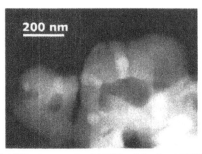

Figure 6. HAADF micrograph of AY95 calcined at 1300°C for 0.5 hr

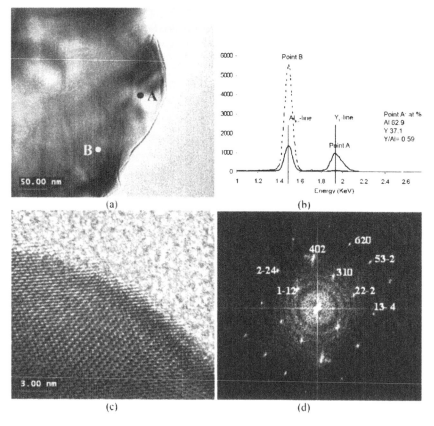

Figure 7. (a): TEM micrograph of a AY95 calcined at 1300°C for 0.5 hr;
(b): EDX peaks intensities of Y and Al determined on points A and B of Figure (a); (c): HR-TEM image of the same particle and (d): indexed Fast Fourier Transfer of image (c)

This point has been confirmed by HR-TEM image acquisition close to point A zone. A higher magnification image of the same particle is shown in Figure 7 c and the corresponding FFT pattern, reported in Figure 7 d. is then indexed according to lattice distances and respective angles between lattice planes, given by JCPDS file 33-40 for $Al_5Y_3O_{12}$ phase.

The influence of the yttrium doping on the alumina sintering behaviour was then investigated. NanoTek powder was dispersed under magnetic stirring; the slurry was divided into two batches, one of which was doped, and then both dried in a oven. The doped powder was also calcined at 600°C for 0.1 hr. Both powders were pressed in bars and then sintered into a dilatometer up to 1500°C for 3 hrs and their densification curves are compared in Figure 8.

AY95 shows a significant increase of the onset sintering temperature (from about 1040°C to about 1150°C), as already reported in literature[27,31,37] for yttrium-doped aluminas. A two-steps sintering behaviour was observed for both samples, mostly imputable to the phase transformation from transition to alpha-alumina and to α-phase sintering, respectively[38,39]. The peak on the derivative curve corresponds to the temperature of maximum rate of transformation phenomena, during the first sintering step and it is located at 1145°C and 1227°C for pure and doped samples, respectively.

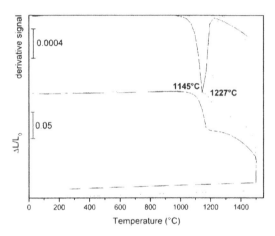

Figure 8. Dilatometric and derivative curves of pure NanoTek powder (solid line) and of AY95 (dashed line) up to 1500°C for 3 hrs.

In addition, the influence of the AY95 pre-treatment temperature on the sintering behaviour was investigated. AY95 powder, calcined at 900°C and 1150°C for 0.5 hr, were uniaxially pressed in bars and sintered under the same conditions of the above 600°C-calcined sample.

The green densities were affected by the calcination temperature, since the values of 1.98, 1.75 and 1.69 g/cm^3 were measured for samples pre-treated at 600°, 900°C and 1150°C, respectively. These differences could be imputed to a less effective particles arrangement in the higher-temperatures treated powders during pressing, probably induced by a more relevant hard agglomeration with increasing the pre-treatment temperature. After sintering, the total linear shrinkage of the 600°C-calcined powder was 19.8%, while a higher value (20.9%) was recorded for the higher-temperature treated materials. To calculate the final fired densities, it was also considered that the three materials underwent different residual mass loss during firing up to 1500°C. In particular, mass loss of 2.5, 1.8 and 1.2% were recorded for samples pre-treated at 600°C, 900°C and 1150°C, respectively, yielding final densities of 93.9%, 93.3% and 91.3% of the theoretical value (3.99 g/cm³ as calculated for the AY95 material by applying the rule of the mixture).

In addition, the role of the heating rate during sintering was investigated by performing dilatometric measurements at a low heating rate (1°C/min). In order to better investigate the whole temperature range, from 700°C to 1500°C, in which the main phenomena involved during thermal treatments take place, such study was carried out on the 600°C-treated powder (Figure 9).

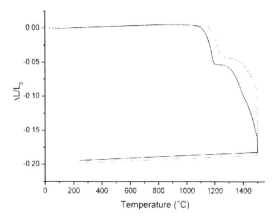

Figure 9. Dilatometric curves of AY95 sintered up to 1500°C for 3 hrs (heating rates: 10°C/min for both materials up to 700°C; then 10 °C/min (dotted line) and 1°C/min (solid line) in the 700°-1500°C temperature range)

Both samples underwent a similar total linear shrinkage of about 19.5% during sintering, reaching a final density of about 95% of the theoretical value. However, the lowering of the onset temperature (at about 1100°C) as well as a more relevant shrinkage during the 2nd shrinkage step were clearly observed in the material sintered at 1°C/min.

To reduce the time of dispersion of the NanoTek powders prior to doping, a ball-milling procedure was set-up in order to achieve a comparable particle size distribution. The above goal was achieved by ball-milling the aqueous suspension (33 wt.% of solid content) for 2.5 hrs by using a powder/spheres weight ratio of 1/10. The dilatometric curve of a pressed bar of the above powder, recorded at a heating rate of 1°C/min in the temperature range 700°-1500°C, is presented in Figure 10 (solid line), and compared to that of a magnetically stirred doped sample (dashed line). The ball-milled material presents an increased linear shrinkage, which is almost completed during the heating step. The final density of the ball milled material was about 3.97 g/cm³, corresponding to 99.5 % of the theoretical value; its derivative curve, reported in the same figure, allows to evidence that the maximum sintering rate occurred at about 1425°C. As a consequence, a ball-milled, doped material was also sintered up to 1450°C for 3 hrs, reaching a final density of about 95% of the theoretical value.

Selected AY95 sintered samples were submitted to SEM/ESEM observations. In Figure 11a, the microstructure of doped-NanoTek, sintered at 1500°C for 3 hrs, is reported, showing a highly dense microstructure, made of well-facetted alumina grains of about 1 µm in size. The BSE-SEM image in the insert allows to observe the YAG particles, having a mean grain size of about 500 nm, homogeneously located in the alumina matrix, mostly at inter-granular positions. The lowering of the maximum sintering temperature to 1450°C for 3 hrs, resulted in a refinement of the above microstructure (Figure 11 b), in spite of a slight decrease of the fired density. The BSE-SEM image shows a highly homogeneous microstructure, made of well-facetted alumina grains of about 1 μm in size and of a well-distributed equiaxial YAG particles with a mean size of about 300 nm.

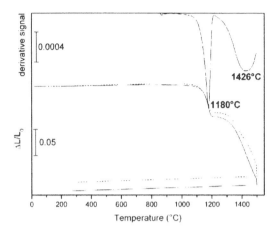

Figure 10. Dilatometric curves up to 1500°C for 3 hrs of AY95 (heating rate 1°C/min in the range 700°C-1500°). NanoTek powder dispersed under magnetic stirring (dashed line) and by ball milling (solid line)

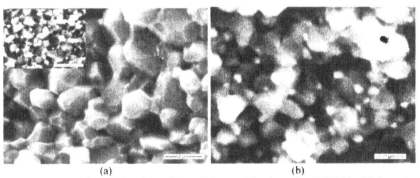

(a)   (b)

Figure 11. SEM micrographs of doped-NanoTek materials, sintered at 1500°C (a, SE-image) and at 1450°C (b, BSE-image) for 3 hrs. The insert of Figure 11 a is a BSE-image.

CONCLUSIONS

In this work, the elaboration of micro-nano 95 vol.% Alumina - 5 vol.% YAG composite materials is presented, by applying an yttrum-doping procedure to a nanocrystalline transition alumina.

The doped powder were fully characterized in terms of phase development by XRD, morphological and crystallographic evolution by HR-TEM and also the dopant distribution as a function of calcination treatment on the alumina particles was monitored by EDX analysis.

Pressureless sintering was used to consolidate the above powders up to full densification.

The final goal, that is the preparation of nano-structured composites in which YAG grains, with a mean size of 200-300 nm, are uniformly dispersed in a fine-grained alumina matrix, was achieved thanks to the selection of an optimised procedure.

In particular, the dispersion process, the pre-treatment temperature and the heating rate during sintering were modified to improve the final microstructure. The best result was achieved by coupling a ball-milling step, performed on pure-alumina, to a low heating rate during sintering, thus allowing to produce almost fully dense, fine-grained sintered bodies.

ACKNOWLEDGMENTS

The Authors wish to thank the European Commission and the Italian Ministry MIUR to have partially supported this research in the framework of the IP project NANOKER and of the project Legge 449/97 "Materiali compositi per applicazioni strutturali di rilevante interesse industriale", respectively.

REFERENCES

[1] P. Harmer, H.M. Chan, G.A. Miller, Unique opportunities for microstructural engineering with duplex and laminar ceramic composites, J. Amer. Ceram. Soc. **75**, 1715-1728 (1992)

[2] A.G. Evans, Perspective on the development of high-toughness ceramics, J. Amer. Ceram. Soc. **73**, 187-206 (1990)

[3] B.K. Jang, M. Enoki, T. Kishi, H.K. Oh, Effect of second phase on mechanical properties and toughening of Al_2O_3 based ceramic composite, Composites Engineering, **5**, 1275-1286 (1995).

[4] M. Sternitzke: Review, Structural ceramic nanocomposites, J. Europ. Ceram. Soc. **17**, 1061-1082 (1997).

[5] F.F. Lange, M.M. Hirlinger, Hindrance of grain growth in Al_2O_3 by ZrO_2 inclusions, J. Amer. Ceram. Soc. **67**, 164-168 (1984).

[6] J.D. French, M.P. Harmer, H.M. Chan, G.A. Miller, Coarsening-resistant dual-phase interpenetrating microstructure, J. Amer. Ceram. Soc. **73**, 2508-2510 (1990).

[7] S. Lartigue and L.Prister, Dislocation activity and differences between tensile and compressive creep of yttria doped alumina, Materials Science and Engineering A **164**, 211-215 (1993).

[8] J.D. French, J. Zhao, M.P. Harmer, H.M. Chan, G.A. Milller, Creep of duplex microstructures, J. Amer. Ceram. Soc.**77**, 2857-2865 (1994)

[9] Q. Tai, A. Mocellin: Review, High temperature deformation of Al_2O_3-based ceramic particle or whisker composites, Ceramics International **25**, 395-408 (1999).

[10] T. Ohji, Creep inhibition of ceramic/ceramic nanocomposites, Scripta Materialia, **44**, 2083-2086 (2001).

[11] R. Torrecillas, M. Schehl, L.A. Diaz, J.L. Menendez, J.S. Moya, Creep behaviour of alumina/YAG nanocomposites obtained by a colloidal processing route, J. Europ. Ceram. Soc. **27**, 143-150 (2007).

[12] L.N. Sataptyh A.H. Chokshi, Microstructural development and creep deformation in an alumina-5% yttrium aluminium garnet composite, J. Amer. Ceram. Soc. **88**, 2848-2854 (2005).

[13] H. Duong, J. Wolfestine, Creep behaviour of fine-grained two-phase Al_2O_3-$Y_3Al_5O_{12}$ materials: Materials Science and Engineering A **172**, 173-179 (1993).

[14] S.M. Choi, H. Awaji, Nanocomposites-a new material design concept, Science and Technology od Advanec Materisl **6**, 2-10 (2005).

[15] Y.K. Jeang, K. Niihara, Microstructure and mechanical properties of pressureless sintered Al_2O_3/SiC nanocomposites, NanoStructured Materials, **9**, 193-196 (1997).

[16] B. Derby, Ceramic nanocomposites: mechanical properties, Current Opinion in Solid Sate and Materials Science, **3**, 490-495 (1998).

[17] D.B. Marshall, J.E. Ritter, Reliability of advanced structural ceramics and ceramic matrix composites-A review, Ceramic Bullettin, **66**, 309-317 (1987)

[18] J.D. Kuntz, G.D. Zhan, A.K. Mukherjee, Nanocrystalline-matrix ceramic composites for improved fracture toughness, MRS Bulletin, 22-27 (2004)

[19] J.D. French, H.M. Chan, M.P. Harmer, G.A. Miller, High-temperature fracture toughness of duplex microstructures, J. Amer. Ceram. Soc. **79**, 58-64 (1996).

[20] H. Wang, L. Gao, Z. Shen, M. Nygren. Mechanical properties of Al_2O_3-5 vol.% YAG composites, J. Europ. Ceram. Soc. **21**, 779-783 (2001).

[21] A. Towata, H.J. Hwang, M. Yasuoka, M. Sando, K. Niihara, Preparation of polycrystalline YAG/alumina composite fibers and YAG fiber by sol-gel method, Composites: Part A **32**, 1127-1131 (2001)

[22] K. Okada, T. Motohashi, Y. Kameshima, A. Yasumori, Sol-gel synthesis of YAG/Al_2O_3 long fibres from water solvent system, J. Eur. Cer. Soc. **20**, 561-567(2000)

[23] W.Q. Li, L. Gao, Processing, microstructure and mechanical properties of 25 vol% YAG-Al_2O_3 nanocomposites, NanoStructured Materials, **11**, 1073-1080 (1999).

[24] H. Wang, L. Gao, Preparation and microstructure of polycrystalline Al_2O_3-YAG composites, Ceramics International **27**, 721-723 (2001).

[25] P. Palmero, A. Simone, C. Esnouf, G. Fantozzi, L. Montanaro,: Comparison Among Different Sintering Routes for Preparing Alumina-YAG Nanocomposites, J. Eur. Cer. Soc. **26**, 941-947 (2006).

[26] M. Schehl, L.A. Díaz, and R. Torrecillas, Alumina Nanocomposites from Powder-Alkoxide Mixtures, Acta Materialia, **50**, 1125-1139 (2002)

[27] R. Voytovych, I. Mac Laren, M.A.Gülgün, R.M.Cannon and M. Rühle, The Effect of Yttrium on Densification and Grain Growth in α-Alumina, Acta Materialia, **50**, 3453-3463 (2002)

[28] M. K. Loudjani and C. Haut, Influence of the oxygen pressure on the chemical state of yttrium in polycrystalline α-alumina. Relation with microstructure and mechanical toughness, Journal of the European Ceramic Society, **16**, 1099-1106 (1996)

[29] P. Gruffel and C. Carry, Effect of grain size of yttrium grain boundary segregation in fine-grained alumina, Journal of the European Ceramic Society, **11**, 189-199 (1993).

[30] R.C. McCune, W.T. Donlon, and R.C. Ku, Yttrium segregation and YAG precipitartion at surfaces of yttrium-doped α-Al_2O_3.

[31] M.K. Cinibulk, Effect of yttria and yttrium-aluminum garnet on densification and grain growth of alumina at 1200°-1300°C, Journal of the American Ceramic Society, 2004, **87**, 692-695.

[32] J. D. Cawley, J.W. Halloran, Dopant distribution in nominally yttrium-doped sapphire, Journal of the American Ceramic Society, **69**, C-195-C-196 (1986).

[33] M. K. Loudjani, J. Foy and C. Haut, Study by extended X-ray absorption fine-structure technique and microscopy of the chemical state of yttrium in α-polycrystalline alumina, Journal of the American Ceramic Society, **68**, 559-562, 1985

[34] http://www.nanophase.com

[35] D. Fargeot, D. Mercurio and A. Dauger, Structural characterization of alumina metastable phases in plasma sprayed deposits", Materials Chemistry and Physics, **24**, 299 – 314 (1990)

[36] H. Ferkel and R. J. Hellmig, Effect of nanopowder deagglomeration on the densities of nanocrystalline ceramic green bodies and their sintering behaviour, NanoStructured Materials, **11**, 5, 617-622 (1999)

[37] J. Fang, A.M. Thompson, M.P.Harmer and H.M. Chan, Effect of yttrium and lanthanum on the final-stage sintering behaviour of ultrahigh-purity alumina, Journal of the American Ceramic Society, **80**, 2005-2012 (1997)

[38] C. Legros, C. Carry, P. Bowen, H. Hoffmann, Sintering of a transition alumina: Effect of Phase Transformation, Powder Characteristics and Thermal Cycle, J. Europ. Ceram. Soc. **19**, 1967-1978 (1999)

[39] P. Bowen, C. Carry, H. Hofmann and C. Legros, Phase transformation and sintering of γ-Al_2O_3 - effect of powder characteristics and dopants (Mg or Y), Key Engineering Materials, 132-136, 904-907 (1997)

NANOSCALE PINNING MEDIA IN BULK MELT-TEXTURED HIGH-T_c SUPERCONDUCTORS AND THEIR IMPORTANCE FOR SUPER-MAGNET APPLICATIONS

M. Muralidhar[1], N. Sakai[1], M. Jirsa[2], M. Murakami[3], I. Hirabayashi[1]

[1]Superconductivity Research Laboratory (SRL), International Superconductivity Technology Center, 1-10-13, Shinonome, Koto-Ku, Tokyo, 135-0062, JAPAN
[2]Institute of Physics, ASCR, CZ-182 21 Praha 8, Czech Republic
[3]Shibaura Institute of Technology, Shibaura 3-9-14, Minato-ku, Tokyo 108-8548, JAPAN

ABSTRACT

The creation and control of nanoscale pinning media in melt-processed ternary superconductors (LRE_1, LRE_2, LRE_3)-$Ba_2Cu_3O_y$ (LRE=light rare earth) represent an important issue not only for the bulk form of these superconductors. Some of the nanometre-size defects are able to increase irreversibility field up to 14 T at 77 K, others extend the pinning performance of the material up to vicinity of critical temperature, enabling thus levitation with liquid oxygen or argon cooling. Additionally, an optimum content of MoO_3 doubles the self-field super-current at 77 K, H//c-axis. Altogether, the pinning tailoring in ternary LRE-123 materials provides a flexible and reliable way to fit the electromagnetic performance with the needs of sophisticated high-temperature and high-magnetic-field applications. The research aims at such applications like super-magnets for NMR, drug delivery systems, wind power generation applications, etc. The recent experimental results on ternary LRE-$Ba_2Cu_3O_y$ materials are reviewed with a special respect to creation and control of novel types of nanoscale pinning media.

INTRODUCTION

Use of high-T_c superconductors in technical applications has the principal attractiveness in the relatively cheap and technically simple liquid nitrogen cooling. For power applications, a high critical current density (J_c) and irreversibility field (H_{irr}) at liquid nitrogen temperature, 77 K, are the most important issues.[1] For the compounds of the rare-earth family, $REBa_2Cu_3O_y$ "RE-123", the J_c and H_{irr} values of Y-123 at 77 K usually serve as a reference as a vast majority of studies in RE-123 has been done on Y-123. However, light rare earth sub-class of the RE-123 materials, LRE-$Ba_2Cu_3O_y$ "LRE-123" (LRE=Nd,Eu,Gd,Sm), has proved to exhibit significantly better electromagnetic properties than Y-123, due to LRE/Ba solid solution, an additional source of point-like pinning sites.

All bulk RE-123 materials distinguish themselves by the ability to trap high magnetic fields, which classifies them to become a new generation of magnets, by order of magnitude more powerful than the best classical permanent magnets.[2,3] In a close future the superconducting super-magnets will enter the scene in NMR, drug delivery systems, water cleaning, wind power plants, and other public applications.[4,5] In all these cases, the high critical current and irreversibility field are the most important issues.

As mentioned above, the $LRE_{1+x}Ba_{2-x}Cu_3O_y$ solid solution clusters "LRE-123ss" play an important role in enhancing flux pinning in a LRE-123 matrix.[6] Their role is similar to that of oxygen-deficient clusters: they are responsible for the formation of the secondary peak on magnetization curve.[7] As all light-rare-earth ions form solid solution with barium, a strong pinning appears also in binary, ternary and quaternary LRE-123 compounds.[8-10] In such mixed compounds a new degree of freedom in pinning tailoring appears, due to variation of the LRE elemental ratio. Different sizes of the LRE atoms introduce a strain disorder in the superconducting lattice, which contributes to flux pinning. In a narrow range of the Nd:Eu:Gd ratio in the (Nd,Eu,Gd)$Ba_2Cu_3O_y$ "NEG-123" compound microstructure analysis revealed that the LRE-123ss clusters arranged into nanoscale planar structures, "lamellas", filling the channels between twin plane boundaries. Due to their thickness and period, comparable to coherence length, these structures represent a very efficient pinning agent, especially at high magnetic fields, leading to a significant enhancement of irreversibility field.[11]

The melt-process technology enables modification of the RE-123 electromagnetic properties by introducing micrometer, sub-micrometer, and even nanometer-sized non-superconducting secondary phase particles, RE_2BaCuO_5 "RE-211". LRE-211 particles refined by ZrO_2 ball milling proved to be particularly effective at low and intermediate magnetic fields. Micro-chemical analysis found that the ball milling led to formation of a new type of Zr-rich ZrBaCuO and (NEG,Zr)BaCuO defects in size of a few tens of nanometers [12]. Efficiency of these particles is inversely proportional to their average size. The effect of Zr in the particle size diminution stacks obviously in Zr chemical inertia in the superconductor matrix. This hypothesis was proved by creating nanoscale particles based on some other inert elements, like MgO,[13] or $Y_2Ba_4CuZrO_y$.[14] We achieved a further substantial improvement in pinning performance of NEG-123 by adding to it nanometer sized MoO_3 particles. In the present paper we report on the microstructure and magnetic properties of these new samples and compare them with the earlier data.

EXPERIMENTAL

In fabrication of ternary LRE-123 superconductors the following technological issues are involved: (i) a proper setting of the matrix chemical ratio; (ii) doping by an appropriate, thoroughly ground secondary phase powder; (iii) formation of a point-like weak pinning disorder of LRE/Ba solid solution clusters via oxygen-controlled melt growth; (iv) optimum oxygenation with respect to the highest T_c. These four tools enabled us to produce materials with the best so far reported parameters.

The samples of $(Nd,Eu,Gd)Ba_2Cu_3O_y$ with various Nd:Eu:Gd ratios, reported in this work, were prepared by a standard melt-growth processing described in Ref. [15]. They were doped with 3 to 40 mol% of $(Nd_{0.33}Eu_{0.33}Gd_{0.33})BaCuO_5$, 0.5 mol% of Pt was added for the secondary phase refinement and 10 wt% of Ag_2O to improve mechanical properties. The pellets were melt-textured under oxygen partial pressure 0.1% O_2 and gas flow rate of 300 ml/min. These samples are compared with $(Nd_{0.33}Eu_{0.33}Gd_{0.33})Ba_2Cu_3O_y$ ones doped by 35 mol% Gd-211 + 1 mol% CeO_2, melt-processed in Ar-1% pO_2 atmosphere.[16] In this case, Gd-211 powder, ball-milled up to 70 nm size, was added before the melt growth process. Newly, nanoscale MoO_3 particles were added in amount of 0.1, 0.2, 0.3, and 0.35 mol% along with the Gd-211 secondary phase particles.

For magnetic measurements small specimens with dimensions of about a x b x c = 2 x 2 x 0.5 mm^3 were cut from the as-grown pellets and annealed in flowing O_2 gas in the temperature range 300-600° C. The microstructure of these samples was studied with a transmission electron microscope (TEM), the atomic force microscopy (AFM), the dynamic force microscope (DFM) and the scanning tunneling microscope (STM). Chemical composition of the matrix was analyzed by energy dispersive X-ray spectroscopy (EDX). Magnetization hysteresis loops (M-H loops) were measured at 77 K using a vibrating sample magnetometer (VSM) with the maximum field of 14 T, parallel to the c-axis. Field sweep rate was 0.6 T/min. The magnetic J_c values were estimated based on the extended Bean's critical state model.[17]

RESULTS AND DISCUSSION

The long-standing dream to utilize bulk high-T_c superconducting magnets at liquid oxygen temperature (90.2 K) has recently become much more feasible due to the progress in fabrication of NEG-123 materials. Although YBa_2Cu_3O, commonly used for levitation at 77 K, has critical temperature (91-93 K),[18] lying only slightly below that of NEG-123, the pinning performance of Y-123 rapidly drops at high temperatures and is therefore insufficient for levitation at 90 K. An even worse situation in this aspect is with BiSrCaCuO, TlBaCuO and other superconductors exhibiting T_c above 100 K.[19,20] These compounds cannot be used for levitation even with liquid nitrogen cooling, due to a low-lying irreversibility line. In present only the new NEG-123 composites exhibit a sufficiently good pinning performance up to 90 K. Fig. 1 shows the field dependence of the super-current density, J_c, for

NEG-123 samples with the Gd-211content ranging between 10 and 50mol%. The $J_c(B)$ curves were deduced from SQUID magnetometer measurements at 77 K in magnetic field applied parallel to c-axis. In all the samples, the initial average secondary phase particles size was <70 nm. It is evident that the super-current density, both at low and high magnetic fields, depends on the amount of initially added secondary phase. The sample with 40 mol% Gd-211 showed the remnant J_c value around 194 kA/cm^2 and J_c at 3 Tesla reached 110 kA/cm^2. This result is by more than 60% better than the previous record values of NEG-123 and by more than order of magnitude better than in other RE-123 materials. With further increase of the Gd-211 content, to 50 mol%, the super-current density already decreased. This may be due to an excessive Zr contamination of the Gd-211 secondary phase. Recently, we found that addition of 0.1mol% nanometer-scale MoO$_3$ particles to the NEG-123 system with 35 mol% of Gd-211 further improved the material performance at 77 K (Fig. 2). In all these samples the initial average Gd-211 particle size was around 70 nm. The low-field super-current density was almost twice as high as in the sample with 0.1 mol% of MoO$_3$ addition. The remnant super-current and H_{irr} decreased with a further increase of the MoO$_3$ content. This J_c value is the highest reported so far for bulk RE-123 materials at 77 K and $H//c$-axis.

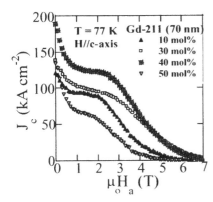

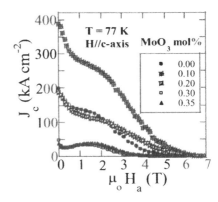

Fig. 1. Field dependence of the super-current density in NEG-123 samples with 10 to 50 mol% Gd-211 refined by ball milling for 4 h (70 nm). All the samples were measured at T = 77 K with $H//c$-axis. The current density increased in the whole field range with increasing content of the secondary phase up to 40 mol% but dropped above this content. Record critical current densities of 194 and 110 kA/cm^2 were achieved at 0 and 3 Tesla, respectively.

Fig. 2. Field dependence of the super-current density in NEG-123 samples with the same, 35 mol% content of Gd-211 (70 nm) but various contents of MoO$_3$. All the samples were measured at T = 77 K with $H//c$-axis. The current density increased in the whole field range up to the 0.1 mol% content of MoO$_3$ and decreased thereafter. Critical current densities as high as 390 and 200 kA/cm^2 were achieved at 0 and 3 Tesla, respectively.

The addition of 30 to 35 mol% of Gd-211 nanoparticles together with a small quantity of ZrO$_2$ or MoO$_3$ to the NEG-123 system not only improved flux pinning at liquid nitrogen temperature but also dramatically enhanced critical currents at higher temperatures, up to vicinity of T_c. As a result, very high critical currents, close to 40 kA/cm^2 at 90 K, were obtained in a sample with 30 mol% Gd-211

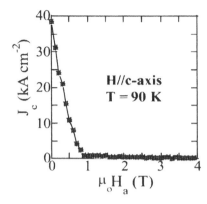

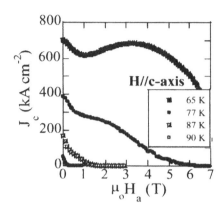

FIG. 3. Field dependence of the critical current density in a (Nd,Eu,Gd)Ba$_2$Cu$_3$O$_y$ sample with 30 mol% Gd-211 refined by ZrO$_2$ ball milling for 4 h (70 nm), measured at $T = 90$ K with $H_a\|c$-axis.

FIG. 4. Field dependence of the critical current density in NEG-123 samples with 35 mol% Gd-211 (70 nm) and 0.1 mol% MoO$_3$ under liquid nitrogen pumping (65 K), at liquid nitrogen (77 K), liquid argon (87K), and liquid oxygen (90.2 K) ($H_a\|c$-axis). Note the very high critical current density of 700 kA/cm^2 at 0 and 4.5 T at 65 K.

nanoparticles (Fig. 3). We can compare this result with the $J_c(H_a)$ data obtained in the sample with 0.1 mol% MoO$_3$ (Fig. 4), detected in the temperature range 65 to 90 K. At 65 K tremendous super-currents were achieved, reaching more than 700 kA/cm^2 at 0 and 4.5 Tesla and exceeding 610 kA/cm^2 over

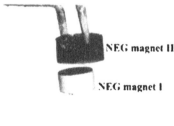

FIG. 5. (left) Permanent Fe-Nd-B magnet levitating above an NEG-123 + 40 mol% Gd-211 (average particle size 70 nm) superconductor at liquid oxygen temperature (90 K), after the superconductor had been magnetized by stray field of the permanent magnet and cooled down to 90 K; (right) NEG-123 pellet suspended below another NEG-123 "permanent" magnet.

the whole field range up to 5 Tesla. These values come close to those of thin films. We note that this achievement might be useful also for technology of thick films of coated conductors. Remnant super-current densities of 175 kA/cm^2 and 50 kA/cm^2 were recorded at liquid argon (87 K) and liquid oxygen (90.2 K), respectively. These results indicate that control of the material properties on the nanometer scale is very important for optimization of the material pinning performance, especially at high temperatures. The levitation and suspension experiments at liquid oxygen temperature 90.3 K are shown in Fig. 5.

Morphology and dispersion of the secondary phase particles in the NEG-123 matrix was studied on several samples by scanning electron microscopy. The data show a uniform dispersion of sub-micrometer sized particles. Our experience says that such a particle size is too large for the observed significant effect on flux pinning at high temperatures. A better insight into the microstructure of this sample was obtained by TEM (Fig. 6). Three types of defects were recognized: large irregular inclusions of about 300 to 500 nm in size, round particles of 20-50 nm (marked by black arrows) and bundles of particles less than 10 nm, marked by white arrows. The chemical composition of the precipitates was studied by scanning TEM-EDX analysis. The analyzed spot of 2-3 nm in diameter enabled to unambiguously analyze even the smallest clusters. The quantitative analysis clarified that the large particles were Gd-211/Gd-rich-NEG-211, in agreement with our earlier studies of the NEG-123 system. In contrast, the defects of the size below 50 nm, marked in Fig. 6 by black arrows, always contained a significant amount of Mo. For the particles less than 5 nm, marked by the white arrows, it was difficult to estimate the exact composition. Anyway, these particles were found to be very effective pins at high temperatures and we succeeded in finding the appropriate processing parameters for their creation. The pinning enhancement due to the new type of defects is so profound that it extends up to temperatures above 90 K. This means that the limiting operating temperature for levitation experiments and other applications shifts from liquid nitrogen (77.3 K) to liquid oxygen (90.2 K) temperature.

FIG. 6. Two transmission electron micrographs of a NEG-123 sample with 35 mol% Gd-211 (average particle size 70 nm) and 0.1 mol% MoO$_3$; the white (left) and black (right) arrows point to some of the nanometer-size Mo-rich particles.

Bulk superconducting super-magnets were recently presented in sputtering apparatuses, in a nuclear magnetic resonance (NMR), in water purification systems and so on. In all these applications use of the super-magnets led to a drastic downsizing and weight reduction compared to the conventional equipments. The super-magnets yield stronger magnetic fields (> 3 T) than the conventional permanent magnets and enable thus qualitative upgrade of the systems features. Small-grain super-magnets were successfully tested on pigs for drug delivery.

While strong trapped magnetic fields (>3 T) were needed in all these applications, the above-described flux pinning enhancement in the NEG-123 system by nanometer-scale secondary phase, Zr or Mo additives was effective in particular at low fields and high temperatures. The ternary composites can be, however, utilized for high field applications, too. Magnetic studies of the NEG-123 system with a varying LRE chemical ratio in the 123 matrix showed a narrow range, where the irreversibility line significantly shifts up-wards. This data indicated that the optimum configuration of Nd:Eu:Gd in the NEG-123 system does not need to be necessarily 1:1:1 and that a variation in the chemical ratio can represent an additional tool in tailoring the pinning landscape in these complex composites. It was also found that each particular Nd:Eu:Gd chemical ratio requires a corresponding optimal concentration of the secondary phase particles [16]. A systematic study of this system finally led to the conclusion that the optimum ratio in the NEG-123 system lies around Nd:Eu:Gd=33:38:28 and the corresponding optimum NEG-211 doping around 5 mol%, which provides irreversibility field above 14 T (Fig. 7).

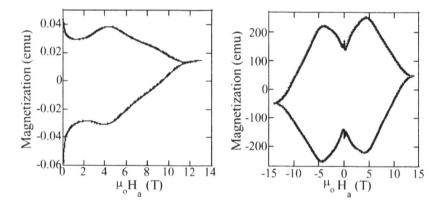

FIG. 7. Magnetic hysteresis loops for the composite (Nd$_{0.33}$Eu$_{0.38}$Gd$_{0.28}$)Ba$_2$Cu$_3$O$_y$ with 3 mol% NEG-211 (left), and 5 mol% NEG-211 (right). Both samples contained 0.5 mol% Pt and were measured at 77 K and H$_a$//c-axis.

Transmission electron microscopy observation of the NEG-123 samples exhibiting high irreversibility (Fig. 7) revealed a structural modulation on a nanometer scale and nanometer-scale lamellar structures.[21] These lamellas were aligned with regular twin boundaries representing thus a fine sub-structure, sometimes straight, sometimes wavy. It had a twin-like structure, but the spacing was much finer, in the order of a few nanometers. Dynamic force microscopy verified the structural features observed by TEM. A similar type of microstructure was also observed in the sample with 7 mol% of NEG-211 and the irreversibility field was in this case above 12 T at 77 K, H//c-axis. On the other hand, no such nanostructures were observed in samples with 30 and 40 mol% NEG-211. Electromagnetic performance of the latter samples at 77 K was nearly equal to that of (Nd$_{0.33}$Eu$_{0.33}$Gd$_{0.33}$)Ba$_2$Cu$_3$O$_y$. Thus, magnetization data and DFM results implied that the high irreversibility at 77 K originated from the observed nanometer-scale microstructure.

A detailed look at the structure by scanning tunneling microscope (STM) (Fig. 8) verified the lamellas linearity and parallelism and dependence of their typical spacing on the initial compound composition. The image of a (Nd$_{0.33}$Eu$_{0.38}$Gd$_{0.28}$)Ba$_2$Cu$_3$O$_y$ sample showed in Fig. 8 exhibited the

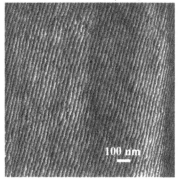

FIG. 8. High magnification STM image of the cleaved surface of (Nd$_{0.33}$Eu$_{0.38}$Gd$_{0.28}$)Ba$_2$Cu$_3$O$_y$ sample. Note the nanoscale lamellas with the average period around 25 nm.

lamellas period around 20-30 nm. In a similar sample doped by 5 mol% NEG-211 the lamellas period dropped to only 3-4 nm (Fig. 9). This size is close to the coherence length in high-T$_c$ compounds and thus close to the vortex core dimensions. This means that an ideal pin size was reached. Chemical analysis across the lamellas clarified that there is a chemical fluctuation in the (Nd+Eu+Gd)/Ba ratio on a nanometer scale. Higher magnification revealed that this structure consists of rows of aligned clusters of non-stoichiometric composition, each cluster being 3 to 4 nm in size (Fig. 9). The tunneling spectra taken on the RE-rich clusters and the regular matrix showed a similar conductivity of both parts. So, both are superconducting regions but T$_c$ is less in the RE-rich clusters. At high magnetic fields the latter become normal and act as a pinning centers (Fig. 9, left). Energy dispersive X-ray analysis "EDX" determined composition of these clusters as (NEG)$_{1.015}$Ba$_{1.985}$Cu$_3$O$_y$. It is important to note that in other ternary LRE-123 materials such compositional fluctuation has been observed, too, but not in the lamellar form. Up to date, such samples showed irreversibility field at 77 K less than 7 T. In NEG-123 with 40 mol% NEG-211 we observed only individual non-correlated clusters of 3-5 nm size (Fig. 10).

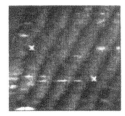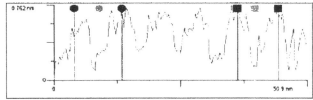

FIG. 9. High magnification STM image of the cleaved surface of (Nd$_{0.33}$Eu$_{0.38}$Gd$_{0.28}$) Ba$_2$Cu$_3$O$_y$ sample with 5 mol% NEG-211 (left). The image size is 50 x 50 nm^2. The tunneling current spectra taken on the RE-rich clusters (read & blue spots) and regular matrix (green spots) showed a similar conductivity of both parts (right figure)

FIG. 10. High magnification STM image of the cleaved surface of a $(Nd_{0.33}Eu_{0.38}Gd_{0.28})Ba_2Cu_3O_y$ sample with 40 mol% NEG-211.

Our study of the ternary LRE-123 systems showed a clear correlation between nanoscale lamellas period and the irreversibility field. The combination of an optimum matrix chemical ratio with an appropriate quantity of a secondary phase, substantial for the formation of the specific matrix structure supporting the high irreversibility field, is quite simple and reliable control tool suitable for industrial applications.

The pinning mechanism associated with the lamellar structure can be explained in the following way: A pin defect extended at least in one dimension along the field direction (columnar track, void, twin plane etc.) is naturally more effective than a number of randomly distributed point-like pins along a single vortex line, due to much larger effective interaction volume of the pinned vortex. It is also well known that presence of a (regular) twin structure does not necessarily result in an overall critical current enhancement. The channeling effect of twins leads to a redistribution of the current flow and appearance of an additional in-plane anisotropy in the sample.[22] This usually results in a suppression of J_c in the region of the secondary peak. The irreversibility field enhancement in presence of a regular twin structure is commonly observed but is not much significant. The novel nanoscale lamellar structure acts on vortices in a similar manner as the conventional twins. It is probably the much higher density that makes this structure of defects so effective at high magnetic fields. Note that this type of nanoscale lamellar structure cannot in principal appear in RE-123 single crystals with heavy RE ions, where RE/Ba solid solution is not possible. As regards the new Mo-based pinning centers significantly affecting the electromagnetic performance at and above 77 K and allowing safe and stable levitation experiments at 90.2 K, it lies between point-like and "large" normal particles. Therefore, its effect on J_c-B dependence is reflected in both low and intermediate fields (around 2 – 3 T). It has been proved that a further defect size reduction and the associated J_c enhancement can be accomplished by simultaneous ball milling of both the secondary phase and matrix composition powders. In contrast to neutron irradiation that produces defects of a similar size but is hardly usable in mass, fast, and economic production,[23] the present procedure is quite simple and for mass production easily adaptable. Both types of novel pinning centers presented in this paper contributed to formation of a new class of bulk high-T_c materials suitable for bulk superconducting super-magnets.

CONCLUSIONS

In this contribution a new ternary LRE-123 material doped by nanoscale particles of MoO$_3$ is presented. This material offers an exceptionally strong flux pinning at liquid nitrogen temperature (77.3 K), as well as an excellent performance even at liquid oxygen temperature (90.2 K). The electromagnetic performance can be easily and quite precisely controlled by the material processing. Both nanoscale lamellar structures and Mo-based non-correlated nanoparticles can be produced. Both new structures are effective pinning agents important for an economically feasible mass production of compact superconducting magnets for high magnetic field (>10 T) and high temperature (>80 K) operation for newly proposed industrial applications.

ACKNOWLEDGEMENTS

This work was supported by Grants-in-Aid for Science Research from the Japan Society for the Promotion of Science (JSPS) and by grant No. 202/08/0722 of Grant Agency of the Czech Republic.

REFERENCES

[1] D. Larbalestier, A. Gurevich, D. M. Feldmann, A. Polyanskii, *Nature*, **414**, 368 (2001).

[2] M. Tomita, M. Murakami, *Nature* **421**, 517 (2003).

[3] M. Muralidhar, M. Jirsa, N. Sakai, I. Hirabayashi, M. Murakami, in *Studies of High Temperature Superconductors*, **50**, 229, edited by A. V. Narlikar (Nova Science Publishers, New York) 2006.

[4] H. Hayashi, K. Tsutsumi, N. Saho, N. Nishizima, K. Asano, *Physica C* **392-396**, 745 (2003).

[5] N. Sakai, S. Nariki, K. Nagashima, T. Miyazaki, M. Muralidhar, I. Hirabayashi, to be published in *Proceeding of the 8th European Conference on Applied Superconductivity*, Brussels, Belgium, 2007.

[6] T. Egi, J. G. Wen, K. Kurada, N. Koshizuka, S. Tanaka, *Appl. Phys. Letters* **67**, 2406 (1995).

[7] M. Daeumling, J. M. Senutjens, D. C. Larbalestier, *Nature* **346**, 332 (1990).

[8] M. Muralidhar, M. Jirsa, N. Sakai, M. Murakami, *Appl. Phys. Letters* **79**, 3107 (2001).

[9] M. Muralidhar, S. Nariki, M. Jirsa, Y. Wu, M. Murakami, *Appl. Phys. Letters* **80**, 1016 (2002).

[10] M. Muralidhar, N. Sakai, M. Jirsa, N. Koshizuka, M. Murakami, *Appl. Phys. Letters* **85**, 3504 (2004).

[11] M. Muralidhar, N. Sakai, N. Chikumoto, M. Jirsa, T. Machi, M. Nishiyama, Y. Wu, M. Murakami, *Phys. Rev. Letters* **89**, 237001 (2002).

[12] M. Muralidhar, N. Sakai, M. Jirsa, N. Koshizuka, M. Murakami, *Appl. Phys. Letters*. **83** 5005 (2003).

[13] B. Zhao, W. H. Song, X. C. Wu, J. J. Du, Y. P. Sun, H. H. Wen, and Z. X. Zhao, *Physica C* **361**, 283 (2001).

[14] N. H. Babu, E. S. Reddy, D. A. Cardwell, and A. M. Campbell, *Supercond. Sci. Technol.* **16**, L44 (2003).

[15] M. Muralidhar, M. Murakami, K. Segawa, K. Kamada, and T. Saitho, *United States Patent*, Patent Number 6,063,753 (2000).

[16] M. Muralidhar, and M. Murakami, in *Studies of High Temperature Superconductors*, **41**, 105, edited by A. V. Narlikar (Nova Science Publishers, New York) 2002.

[17] D. X. Chen, R. B. Goldfarb, *J. Appl. Phys.* **66**, 2489 (1989).

[18] S. Jin, A. Fastnacht, T. H. Tiefel, R. C. Sherwood, *Phys. Rev.* B **37**, 5828 (1988).

[19] Z. Z. Sheng, A. M. Hermann, *Nature* **33**, 138 (1988).

[20] H. Maeda, Y. Tanaka, M. Fukufomi, T. Asano, *Jpn. J. Appl. Phys.* **27**, L209 (1988).

[21] M. Muralidhar, N. Sakai, M. Jirsa, M. Murakami, & N. Koshizuka, Supercond. Sci. Technol. **17**, 1129 (2004).

[22] M. Jirsa, M. R. Koblischka, T. Higuchi, and M. Murakami, *Phys. Rev.* B **58**, R14771(1998).

[23] A. Umezawa, G. W. Grabtree, J. Z. Liu, H. W. Weber, W. K. Kown, L. H. Nunez, T. J. Moran, C. H. Sowers, H. F. Claus, *Phys. Rev.* B **36**, 7151 (1997).

NOVEL NANO-MATERIAL FOR OPTO-ELECTROCHEMICAL APPLICATION

P. C. Pandey, Dheeraj S. Chauhan
Department of Applied Chemistry
Institute of Technology, Banaras Hindu University Varnasi-221005

ABSTRACT

We report herein the development of few types of ormosil-modified electrodes, derived from Hexacyanoferrate- [(Prussian blue systems (PB-1, PB-2, and PB-3), palladium (Pd-) system, graphite (Gr-) system, gold nanoparticle (AuNPs) system and palladium-gold nanoparticle (Pd-AuNPs)] and ruthenium bipyridyl for opto-electrochemical applications. Four approaches causing manipulation in nano-structured domains, i.e. (a) increase in the molecular size of the components generating nano-structured domains; (b) modulation via chemical reactivity; (c) modulation by non-reactive moieties and known nanoparticles; and (d) modulation by mixed approaches (a-c), all leading to decrease in a nano-structured domains; are described. The results demonstrated that an increase in the size of nano-structured domains or decrease in micro-porous geometry increases the efficiency of electro-catalysis. These ormosil-modified electrodes have been used for spectro-electrochemical measurements if encapsulated redox moieties show sharp electro-chromic behavior. Spectro-electrochemistry of few organic redox materials like tetracyanoquinodimethane (TCNQ), tetrathiafulvalene (TTF), Ferrocene including hexacyanoferrate systems are studied and reported in this communication. Typical application of these ormosil materials in electrocatalysis of hydrogen peroxide and dopamine detection are reported.

I. INTRODUCTION

Selective recognition is the inherent property of a biologically derived catalyst which ultimately introduces humanity in mankind. Chemically derived catalysts lack the selectivity of recognition, thus, requires extensive investigation of fundamental and applied researches to generate similar pattern in artificial systems. The current output provides a little finding on these lines.

The present finding concern to develop a system where the possibility of electron-transport and optical measurements are simultaneously conducted as such information may lead better understanding of the real systems. Some of the redox species are chosen for generating such systems in combination with ormosil matrix-known system for optical applications. As an typical example of potassium hexacyanoferrate (III), [$K_3Fe(CN)_6$], which is a biocompatible mediator and has shown potential application in mediated bio-electrochemical detection of several redox chemical and biochemical systems. The current finding demonstrated the introduction of electrocatalytic sites when this mediator is incorporated into nano-structured domains, as the recent finding on nano-structured materials has now introduced the concept of novel technological source in multiple directions with current emphasis in the area of Analytical chemistry.[1,2] We report some novel ormosil-modified electrodes having differential electrocatalytic efficiency originating from similar basic platform via desired chemical-manipulation routs, thus creating a library of electrocatalytic sites within nano-structured domains. These routes are; (i) *in-situ* conversion of $K_3Fe(CN)_6$ into Prussian blue; (ii) introduction of palladium into nano-structured network; (iii) introduction of noble metal catalyst within the nano-pores of the matrix; (iv) introduction of targeted amount of water leach able components within nano-structured domains; (v) increasing the conductivity of the matrix, and (vi) introduction of ion recognition sites within the matrix. These manipulations lead the development of eight types of ormosil modified electrodes for optoelectrochemical applications. The opto-electrochemistry of few systems is reported. The electrocatalysis of hydrogen peroxide and dopamine are reported on these systems.

Some other organic compounds like tetracyanoquinodimethane (TCNQ), tetrathiafulvalene (TTF) and ferrocene are very well known redox systems and have also proved bio-compatibility. These organic redox systems are chosen for generating such material better useful in opto-electrochemistry and also reported in this communication.

II. EXPERIMENTAL

Construction of Ormosil modified electrodes Eight types of modified electrodes, thereafter referred as Hexacyanoferrate-system; PB-1, PB-2, PB-3 systems, Pd-system, Gr-system, AuNPs-system and Pd-AuNPs-system respectively were prepared using three different alkoxysilanes having reactive organic functionalities. The ormosil matrix was prepared by adding the constituents in the sequential order as mentioned in the Table 1, except for PB-3 where aqueous solution of Prussian blue (made by mixing $K_3Fe(CN)_6$ and $FeSO_4$) was used in place of constituents C and D. For *in situ* conversion of potassium hexacyanoferrate(III) into Prussian blue (PB-1), cyclohexanone was mixed with potassium hexacyanoferrate(III) in the presence 3- aminopropyltrimethoxysilane whereas for the *in situ* conversion of potassium hexacyanoferrate(III) into Prussian blue (PB-2), ferrous sulphate was mixed with potassium hexacyanoferrate(III) in the presence of 3-aminopropyltrimethoxysilane. On the other hand, for the preparation of traditional Prussian blue (PB-3) potassium hexacyanoferrate(III) was separately mixed with ferrous sulphate in absence of 3-aminopropyltrimethoxysilane and resulting blue reaction mixture was mixed with the ormosil precursors. Pd-system uses the interaction of palladium chloride with 3-glycidoxypropyltrimethoxysilane where palladium chloride opens the epoxide ring and get reduced into palladium followed by coordination to alkoxysilane as described earlier.[79, 84] Similarly, known electrocatalytic materials namely graphite particle and gold nanoparticle were coupled to the nano-structured network of ormosil to give Gr-system and AuNPs system respectively. Pd-AuNPs system of ormosil is generated from Pd-system by adding gold nanoparticles. The reaction mixture for each system was homogenized on vertex mixer for 10 minutes and 6 μl of each was layered separately on cleaned ITO electrodes. The various ormosil films were allowed to dry for 8-10 hours at 25-30 ^{0} C.

Table 1. Composition of Ormosil's precursors for preparation of modified electrodes (systems 1→8).

System	A (μl)	B (μl)	C (μl)	D (μl)	E (μl)	F (μl)	G (μl)	H (μl)	I (μl)	J (mg)	K (μl)
Hexacyanoferrate-	15	-	35	30	-	-	2	150.	2.5	-	-
PB-1	15	-	35	30	-	2.5	2	147.5	2.5	-	-
PB-2	15	-	35	30	30	-	2	120.	2.5	-	-
PB-3	15	-	35	*	-	-	2	120	2.5	-	-
Pd-	15	7.5	35	30	30	-	2	120	2.5	-	-
Gr-	15	-	35	30	30	-	2	120	2.5	0.5	-
AuNP-	15	-	35	30	30	-	2	90	2.5	-	30
Pd-AuNP	15	7.5	35	30	30	-	2	120	2.5	0.5	-

A- 3-glycidoxypropyltrimethoxysilane; B- aqueous solution of $PdCl_2$ (1 mg / ml), C- 3-Aminopropyltrimethoxysilane; D- aqueous solution of 10 mM $K_3Fe(CN)_6$; E- aqueous solution of 10 mM $FeSO_4$ F- cyclohexanone; G- 2-(3,4-epoxycyclohexyl)ethyltrimethoxysilane; H- double distilled-deionized water; I- HCl (0.5N); J- graphite powder (Gr. particle size 1-2 μ); K- colloidal gold (AuNPs); * 60 μl of aqueous solution of Prussian blue (made by mixing $K_3Fe(CN)_6$ and $FeSO_4$ 10 mM each).

Opto-electrochemistry of ormosil-modified electrodes.

The above mentioned eight types of ormosil modified electrodes have been efficiently useful in electrocatalysis. However, it is obivous to study the spectroelectrochemistry of such ormosil modified electrodes. Potassium ferricyanide has shown electro-chromism as Fe^{2+}/Fe^{3+} show differential abosorbance and could be justified from spectro-electrochemical measurements which has been undertaken. In order to justify wide application of the present material, incorporation of other redox organic moieties justifying sharp change in color was under taken. For such investigation tetracyanoquinodimethane (TCNQ), Tetrathiafulvalene (TTF), and Ferrocene were chosen to encapsulate within the ormosil film. Such films have been analyzed through spectro-electrochemical measurements. The construction of these films incorporates the composition as shown in table-2.

Table 2. Composition of ormosil precursors and redox materials.

System	A (µl)	B (µl)	C (µl)	D (µl)	E (µl)	F (µl)	G (µl)	H (µl)	I (µl)	J (µl)	K (µl)
Ferri	70	10	60	-	5	240	-	-	-	-	-
Ferri-Pd	70	10	60	-	5	240	10	10	-	-	-
Fc	-	10	-	-	5	300	-	-	70	-	-
Fc-Pd	-	10	-	-	5	280	10	10	70	-	-
TCNQ	-	10	-	-	5	300	-	-	-	70	-
TCNQ-Pd	-	10	-	-	5	280	10	10	-	70	-
TTF	-	10	-	-	5	300	-	-	-	-	70
TTF-Pd	-	10	-	-	5	280	10	10	-	-	70
PB	70	10	60	60	5	180	-	-	-	-	-
PB-Pd	70	10	60	60	5	180	10	10	-	-	-

A = Aminopropyltrimethoxysilane. B = 2-(3,4-epoxycyclohexyl)ethyltrimethoxysilane. C = 50mM Potassium Ferricyanide. D = 50 mM Ferrous Sulphate. E = 0.1M HCl. F = Water. G = Glycidoxypropyltrimethoxysilane. H = Palladium Chloride. I = Ferrocene in Aminopropyltrimethoxysilane. J = TCNQ in Aminopropyltrimethoxysilane. K = TTF in Aminopropyltrimethoxysilane

III. RESULTS AND DISCUSSION

Electrochemical applications of the hexacyanoferrate-system encapsulated within ormosil-modified electrode with special attention of hydrogen peroxide oxidation.

The reports during last decade revealed that nano-structured geometry could be easily generated within organically modified sol-gel glasses (ORMOSIL) under ambient conditions. We have been working on generating such platforms for non-mediated / electrocatalytic oxidation of glucose, NADH and hydrogen peroxide.[20-31] We have further investigated that potassium hexacyanoferrate(III) could be easily encapsulated within the ormosil network and chemical sensitizers like. ion exchanger Nafion* and ion carrier crown ether, present in nano-structured network of ormosil, play an important role in the heterogeneous electrochemical oxidation of hydrogen peroxide.[27-29] Further investigating on these lines we generated the concept of introducing variable electrocatalytic efficiency within the same network.. Some schemes subsequently referred as, Hexacyanoferrate-, PB-, Pd-, Gr-, AuNPs, and Pd-AuNPs for generating these sites within ormosil network are reported in this communication. Accordingly, efforts have been made to generate ormosils with variable nano-structured domains. A series of steps have been undertaken for generating variation in such network, which indeed provided valuable analytical information. The details of such process are given below.

1. In-situ conversion of potassium hexacyanoferrate into Prussian blue. Hexacyanoferrate-system generated the ormosil network in a single step protocol where all ormosil's precursors along with potassium hexacyanoferrate (III) were simultaneously homogenized and allowed to form ormosil film. The Prussian blue system (PB-system) involves a two stage process in ormosil formation (table-1). In stage-1, potassium hexacyanoferrate(III) and cyclohexanone were mixed in the presence of 3-aminopropyltrimethoxysilane by stirring, thus leading to the conversion of potassium hexacyanoferrate(III) into Prussian blue.[27-29] The second stage is the mixing of the resultant Prussian blue solution with other ormosil's precursors (Table-1) for generating PB-1-- system. The concentration of redox system is kept constant in both conditions. There is change in redox behavior which is more apparent from electrocatalytic results on the oxidation of hydrogen peroxide (the results are not shown). The electrochemistry of potassium hexacyanoferrate in homogeneous solution and within ormosil network is recorded in Fig.1. The peak current justifies the role of nano-structured network on the electrochemistry of the same keeping similar concentrations of potassium hexacyanoferrate.

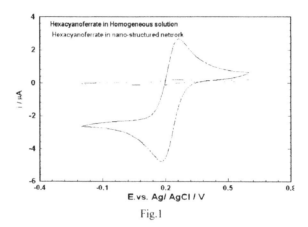

Fig.1

2. Increase in the molecular size / orientation of ormosil's precursors. There is another way of manipulating ormosil network, as compared to that of PB-1-system, by changing the molecular size of one of the reacting components of Prussian blue conversion from potassium hexacyanoferrate (III). Such modification was availed by replacing cyclohexanone with ferrous sulphate, which ultimately changed the size of nano-structured network and resulted in as PB-2 system. Such variation has been recorded by atomic force microscopy of PB-1, PB-2 and PB-3. The AFM images are shown in (Fig. 2). The images show that the nano-geometry of the film increases with decreasing the size of Prussian blue formation-initiating moiety. We further made effort to change the nano-geometry as compared to that of PB-2 system by changing protocol of reaction dynamics basically using same ormosil precursors. Our previous reports[27-29] showed that potassium hexacyanoferrate (III) is converted into Prussian blue by adding cyclohexanone/tetrahydrofuran in the presence of 3-aminopropyltrimethoxysilane. Further, Prussian blue formation is not observed on adding cyclohexanone/tetrahydrofuran when 3-aminopropyltrimethoxysilane was absent or replaced by any other organosilane without having amino-terminated functionality. On the other hand, when ferrous sulphate is added into aqueous solution of potassium hexacyanoferrate (III), Prussian blue formation is triggered both in the absence and in the presence of 3-aminopropyltrimethoxysilane. The presence of 3-aminopropyltrimethoxysilane generated PB-2 system whereas, absence of amino-terminated organosilane during Prussian blue conversion generated PB-3 system with relatively much reduced nano-geometry as shown in Fig.2 (PB-3).

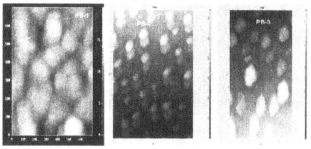

Fig.2

It should also be noted that the particle size of Prussian blue depend on the curing protocol and subsequent mixing of the same with other ormosil precursors for triggering ormosil formation. The AFM images as shown in Fig.2 supported our expectation that gradual increase in size of Prussian blue formation initiating agents decreases the nano-geometry of ormosil network. The variation in electrochemistry of these three systems can be evaluated from Fig.3.

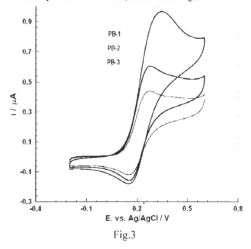

Fig.3

3. *Modulation via chemical reactivity*. Another approach for changing the nano-structured domains in ormosil network is to introduce another reaction system prior to initiating ormosil network formation. Such introduction is availed via chemical reactivity of one of the organosilane participating in ormosil formation. Our previous reports[24, 26] suggested that 3-glycidoxypropyltrimethoxysilane is highly sensitive to the presence of palladium chloride. The epoxide ring of *glymo*-group is opened by palladium chloride followed by the reduction of

Palladium (II) into palladium. The reduced palladium is then coordinated with two moiety of glymo-residue[24]. as given below:

$$(CH_3O)_3Si\ (CH)_3OCH_2CH{-}CH_2$$
$$Pd$$
$$(CH_3O)_3Si\ (CH)_3OCH_2CH{-}CH_2$$

When above reaction product is used for ormosil formation Pd-system is generated. The introduction of palladium within ormosil network not only altered the nano-geometry but also affected the mechanistic approach on the electrochemistry of hydrogen peroxide as well.

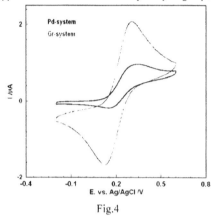

Fig.4

4. Modulation via known nano-particles. The availability of known nano-particles (NPs) compatible to nano-structured domains is now frequently used in electrocatalysis. These moieties are carbon nano-tube and gold nano-particles (AuNPs). Such moieties are now easily converted into suspension and are well-suited to incorporate within the nano-geometry of ormosil film. We started such modulation using graphite micro-particles of particle size 1 μ within the ormosil film and Gr-system was generated. The inclusion of such micro-particle increases the porosity of the ormosil film thus increasing the translational degree of freedom of Prussian blue within ormosil film. The result for the Gr-system is shown in (fig. 4) which again supported the expected trend on hydrogen peroxide oxidation. Another system was made by incorporating gold nano-particle suspension within the ormosil film and AuNPs-system was generated. Incorporation of gold nano-particle provided remarkable finding on the electrochemical behavior as shown in Fig.5.

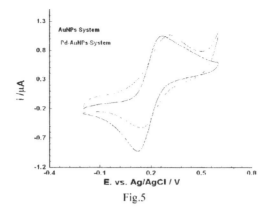

Fig.5

There are two remarkable observations on the electrochemistry of hydrogen peroxide at the surface of AuNPs-system; (i) decrease in overvoltage of hydrogen peroxide oxidation compared to that all other systems and (ii) decrease in the reduction kinetics of hydrogen peroxide.

5. Modulation via mixed protocols. Another method for altering the nano-geometry of ormosil film may be availed by combining mixed approaches as discussed above. Pd-AuNPs-system was developed by combining the approaches adopted for Pd-system (attachment of palladium) along with the incorporation of gold nano-particle (AuNPs). The electrochemistry of the same is shown in Fig.5. The electrocatalytic efficiency of this system is found excellent compared all othere system based on encapsulation of potassium hexacyanoferrate within ormosil network.

We used the above material for electrocatalytic oxidation of hydrogen peroxide. A typical result on prussian blue system is shown in Fig.6. The efficiency of oxidation is better for PB-1 system compared to that of PB-2 system which in turn is better as compared to that of PB-3 system. These results suggest that an increase in nano-structured network increases the electrocatalytic efficiency of the material.

Electrochemical applications of the Ruthenium bipyridyl encapsulated within ormosil-modified electrode with special attention of hydrogen peroxide oxidation.
Ruthenium bipyridyl is known electroluminescent material for technological applications. We have encapsulated the same within the ormosil network originally derived from above reported precursors. The encapsulated material shows excellent electrochemistry as recorded in Fig.7.

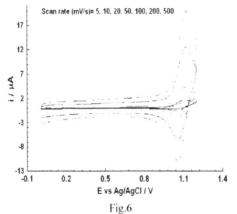

Fig.6

The typical application of the material is electrocatalytic detection of dopamine based on both electrochemical and electroluminescent property of the material. The results on electrocatalytic oxidation are shown in Fig.8.

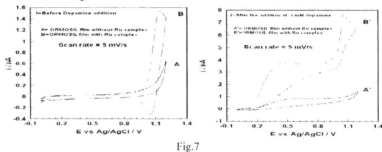

Fig.7

We also coupled ruthenium bipyridyl along with other modulation as discussed for hexacyanoferrate system and got very interesting observation. The results based on spectroelectrochemistry are ongoing and will be reported subsequently.

Spectroelectrochemistry of some novel organic redox material encapsulated ormosil.

We encapsulated tetracyanoquinodimethane (TCNQ). tetrathifulvalene (TTF) and ferrocene encapsulated ormosil modified electrode on ITO electrode. These organic redox species are electrochrmic and could be conveniently used both for efficient electron transport and photon transport. The spectroelectrochemical measurements of these systems including hexacyanoferrate are shown in Fig.9 a, 9b, 9c and 9 d for hexacyanoferrate-palladium system. ferrocene-palladium. TCNQ-palladium. and TTF -palladium systems respectively.

Ferricyanide-Pd

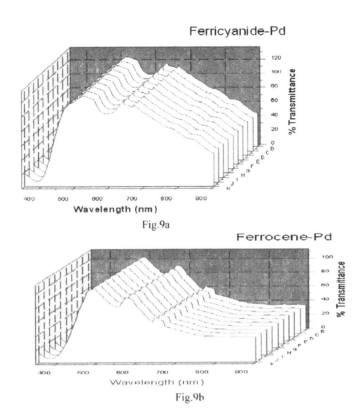

Fig.9a

Ferrocene-Pd

Fig.9b

TCNQ-Pd

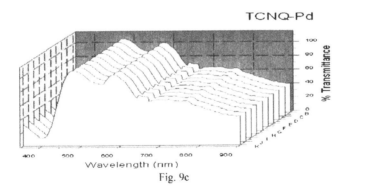

Fig. 9c

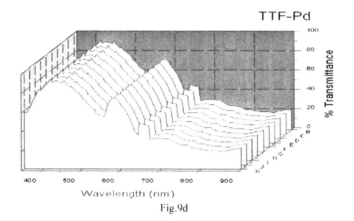

Fig.9d

REFERENCES

[1] A. C. Balazs. T. Emrick and T. P. Russell. *Science*. 314, 1107 (2006).

[2] R. W. Murray. *Anal. Chem.*, 79. 1 (2007).

[3] M. C. Burt and B. C. Dava, *J. Am. Chem. Soc.*. 128, 11750 (2006).

[4] G. Palmisano. E. Le Bourhis. R. Ciriminna, D. Tranchida and M. Pagliaro. *Langmuir*. 22. 11158 (2006).

[5] T. K. Das, I. Khan. D. L. Rousseau and J. M. Friedman. *J. Am. Chem. Soc.*, 120. 10268 (1998).

[6] A. Fidalgo. R. Ciriminna, L. M. Ilharco and M. Pagliaro. *Chem. Mater*. 17. 6686 (2005).

[7] M. Opallo and M. Saczek-Maj. *Chem. Comm.*. 5. 448 (2002).

[8] I. Zawisza. J. Rogalski and M. Opallo. *J. Electroanal. Chem.*, 588, 244 (2006).

[9] P. N. Deepa. M. Kanungo. G. Claycomb, P. M. A. Sherwood and M. M. Collinson, *Anal. Chem.*. 75, 5399 (2003).

[10] Y. Fu. M. M. Collinson and D. A. Higgins. *J. Am. Chem. Soc.*, 126, 13838 (2004).

[11] S. Sayen. M. Etienne. J. Bessiere and A. Walcarius. *Electroanalysis*. 14, 1521 (2002).

[12] A. Walcarius. D. Mandler. J. A. Cox. M. M. Collinson. and O. Lev. *J. Mater. Chem.*, 15. 3663 (2005).

[13] L. M. Ellerby. C. R. Nishida, F. Nishida, S. A. Yamanaka, B. Dunn. J. S. Valentine and J. I. Zink, *Science*. 255, 1113 (1992).

[14] B. C. Dave. B. Dunn. J. S. Valentine and J. I. Zink. *Anal. Chem.*, 66. A1120 (1994).

[15]   F. Nishida, J. M. Mckiernan, B. Dunn, J. I. Zink, C. J.M. Dequaire, B. Limoges, J. Moiroux and J-M. Saveant, *J. Am. Chem. Soc.*, 124, 240 (2002).

[16]   J. D. Qiu, H. Z. Peng, R. P. Liang, J. Li and X. H.Xia, *Langmuir*, 23, 2133 (2007).

[17]   A. A. Karyakin, O. V. Gitelmacher and E. E. Karyakina, *Anal. Chem.*, 67, 2419 (1995).

[18]   F. Ricci and G. Palleschi, *Biosensors Bioelectron.*, 21, 389 (2005).

[19]   A. A. Karyakin, E. A. Puganova, I. A. Budashov, I. N. Kurochkin, E. E. Karyakina, V. A. Levchenko, V. N. Matveyenko and S. D. Varfolomeyev, *Anal. Chem.*, 76, 474 (2004).

[20]   M. R. Miah and T. Ohsaka, *J. Electrochem. Soc.*, 153, E195 (2006).

[21]   P. C. Pandey, S. Upadhyay and H. C. Pathak, *Electroanalysis*, 11, 59 (1999).

[22]   P. C. Pandey, S. Upadhyay, H. C. Pathak and C. M. D. Pandey, *Electroanalysis*, 11, 950 (1999).

[23]   P. C. Pandey, S. Upadhyay, H. C. Pathak, I. Tiwari and V. S. Tripathi, *Electroanalysis*, 11, 1251 (1999).

[24]   P. C. Pandey, S. Upadhyay, I. Tiwari and S. Sharma, *Electroanalysis*, 13, 1519 (2001).

[25]   P. C. Pandey, S. Upadhyay, N. K. Shukla and S. Sharma, *Biosens. Bioelectron.*, 18, 1257 (2003).

[26]   P. C. Pandey, S. Upadhyay and S. Sharma, *J. Electrochem. Soc.*, 150, H85 (2003).

[27]   P. C. Pandey, B. C. Upadhyay and A. K. Upadhyay, *Anal. Chim. Acta*, 523, 219 (2004).

[28]   P. C. Pandey, B. C. Upadhyay and A. K. Upadhyay, *Sens. Actuators B*, 102, 126 (2004).

[29]   P. C. Pandey and B. C. Upadhyay, *Talanta*, 67, 997 (2005).

[30]   P. C. Pandey, B. Singh, R. C. Boro and C. R. Suri, *Sens. Actuators B*, 122, 30 (2007).

[31]   P. C. Pandey, S. Upadhyay, I. Tiwari and V.S. Tripathi, *Sens. Actuators B*, 72, 224 (2001).

[32]   A. A. Karyakin, E. E. Karyakina and L. Gorton. *Anal. Chem.*, 72, 1720 (2000).

[33]   A. A. Karyakin. *Electroanalysis*, 13, 813 (2001).

[34]   A. A. Karyakin, E. A. Kotel'nikova, L. V. Lukachova, E. E. Karyakina and J. Wang, *Biosens. Bioelectron.*, 74, 1597, 2002.

[35]   D. Moscone, D. D'Ottavi, D. Compagnone, G. Palleschi and A. Amine, *Anal. Chem.*, 73, 2529 (2001).

[36]   I. L. de Mattos, L. Gorton and T. Ruzgas, *Biosens. Bioelectron.*, 18,193 (2003).

KAOLINITE-DIMETHYLSULFOXIDE NANOCOMPOSITE PRECURSORS

Jefferson Leixas Capitaneo [1], Valeska da Rocha Caffarena [2], Flavio Teixeira da Silva[3], Magali Silveira Pinho [4], Maria Aparecida Pinheiro dos Santos [4]

[1] Centro Universitário Estadual da Zona Oeste (UEZO)/ Instituto de Macromoléculas, IMA/UFRJ, jeff@ima.ufrj.br[2] Petróleo Brasileiro S.A – Gas&Power - Av. Almirante Barroso, 81 – 31° andar – Centro, ZIP CODE 20031-004, Rio de Janeiro, Brazil, valeskac@petrobras.com.br
[3] Dep. of Metallurgical and Materials Engineering, COPPE-PEMM/UFRJ, PO Box 68505, Zip Code 21941-972, Rio de Janeiro, RJ, Brazil, flatesi@metalmat.ufrj.br
[4] Brazilian Navy Research Institute (IPqM) - Rua Ipiru 2, Praia da Bica, Ilha do Governador, Rio de Janeiro, RJ, Brazil, Zip Code 21931-090, magalipinho@yahoo.com.br, cidarta.cidarta@gmail.com

ABSTRACT

Lamellar kaolinite matrices were subjected to intercalation reactions using dimethylsulfoxide (DMSO) as the intercalating agent. The nanocomposites, kaolinite-DMSO (K-DMSO), were obtained by two distinct routes while the DMSO:H_2O (9% v/v) ratio was kept constant. Aliquots were removed periodically to evaluate the evolution of the intercalation of the DMSO molecules in the kaolinite lamellae, using X-ray diffraction (XRD), Fourier transform infrared spectroscopy (FTIR), scanning electron microscopy (SEM), energy dispersive X-ray spectroscopy (EDS), and thermogravimetric analysis (TGA). The results indicated that the modification of the methodology normally used for DMSO intercalation in kaolinite led to a significant reduction in the time required for intercalation (from 480 h to 5 h) and that the synthesis carried out at 95-100°C produced samples with an 87% intercalation rate.

INTRODUCTION

Micro and macrocomposites have been used in advanced devices in the automotive, space and biomedical industries. However, a greater integration among the different types of materials is restricted to the dimension of the phases involved. In order to maximize this interaction, which means using the potentialities of each material, the number of surfaces and interfaces must be expanded, which can be achieved by using nanocomposites.

These nanomaterials include nanocomposites with clay minerals, which have promising applications in a variety of areas. Clay minerals can be used as two-dimensional matrices (quantum walls) for the insertion of different organic molecules in nanometric spaces existing in their crystalline structure [1 – 3].

It is important to point out that the clay minerals utilized to obtain nanocomposites are generally 2:1 type phyllosilicates (mainly montmorillonite, hectorite and saponite) and not kaolinite, although Brazil is one of the major producers of this clay mineral.

Due to the hydrogen bonds between its lamellae, which are characteristic of type 1:1 phyllosilicates, kaolinite is able to directly intercalate a limited number of small host molecules such as N-methylformamide (NMF), dimethylsulfoxide (DMSO) and nitroanillin [4].

Note that the intercalation ability in kaolinite depends on the size of the intercalating molecules (hosts) and their polarity, i.e., only small organic molecules possessing a high dipolar moment are able to directly intercalate in this clay mineral. This difficulty to intercalate can be explained by the fact that the lamellae of kaolinite are bound to each other by means of hydrogen bridges involving the Al-OH and Si-O groupings. The intercalatable polar molecules are stabilized through dipolar interactions, hydrogen bridges and van der Waals forces [5, 6].

In the intercalation of larger sized species such as polymers, one of the chemical routes employed is based on the Chemical Displacement Method, whereby new species can be intercalated by the displacement of the small molecule inserted previously.

With the use of pre-expanded precursors, the molecules do not encounter all the resistance they would find to break the hydrogen bridges existing between the lamellae of the non-intercalated matrix, and should therefore simply substitute the smaller molecule [4].

Ranking high among intercalating precursor agents is dimethylsulfoxide (DMSO), whose physical properties are listed in Table 1.

Table 1 – Physical Properties of dimethylsulfoxide (DMSO):

DMSO	Physical Properties	
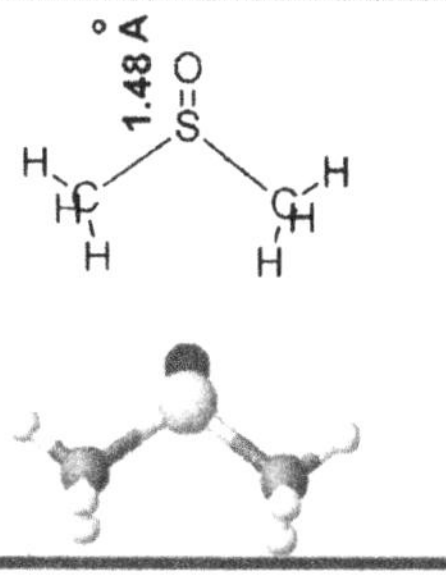	Molecular weight	78,13 g/mol
	Boiling Point	189.0 °C
	Density	1.1004g/ ml at 20 °C
	Dielectric Constant	46.68 at 20 °C
	Dipolar Momentum	4.3 D at 25 °C
	Solubility (water)	Miscible

These physical properties render it a solvent with peculiar characteristics, able to solubilize organic and inorganic compounds that are insoluble in most other liquids. DMSO is an aprotic solvent (it does not have hydrogen atoms bound to atoms of high electronegativity, such as oxygen), highly polar (with a dipole moment of 4.3 D) and a proton receptor. Its molecule is pyramidal, with the corners of the pyramid occupied by carbon, oxygen and sulfur atoms. The –SO bond can be described as a resonance hybrid (Figure 1) between a semipolar double bond (with an estimated dipole moment of 3.0 D) and a type $(p{\rightarrow}d)_\pi$ double bond. This explains the high value of the dipolar moment of this molecule and its relatively high basicity, although the bond length (1.48 Å) is close to the value expected for a double bond [7, 8].

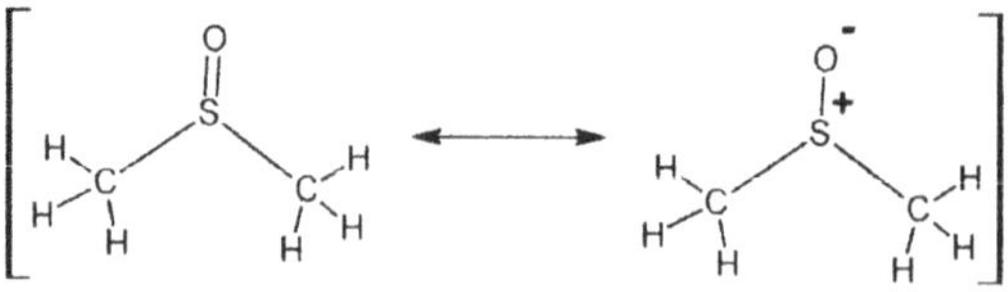

Figure 1: Canonical forms of the resonance hybrid that best represents the DMSO molecule.

When insertion of DMSO molecules occurs between the kaolinite layers, there is a pre-expansion of the interlamellar space, whose basal interplanar distance increases from approximately 0.72 nm to about 1.11 nm (Figure 2).

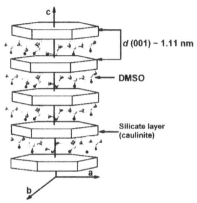

Figure 2: Kaolinite layers intercalated with DMSO molecules.

One of the major obstacles to the use of dimethylsulfoxide (DMSO) as an intercalating agent is the relatively long time required to obtain the kaolinite-DMSO (K-DMSO) precursor, i.e., from 20 days to several months.

One of the first studies involving DMSO intercalation in kaolinite was conducted in 1968 by Olejnik et al. [6]. Their experiments involved putting 20 mg of well-crystallized kaolinite in contact with pure DSMO and DMSO/ H_2O solutions in varied proportions. The suspensions were kept at room temperature for several hours and the evolution of the intercalation was monitored by X-ray diffraction (XRD) and Fourier transform infrared spectroscopy (FTIR). The best results were achieved using a DMSO/H_2O (9% v/v) solution and the water acted as a catalyzer, increasing the speed of the intercalation reaction, a fact that can be explained by the solvation of the DMSO molecules, which increased the interaction between the DMSO and the hydroxyls of the kaolinite structure [6].

Higher percentages of water do not favor the intercalation process of DMSO, for the DMSO molecules become highly solvated, leaving few molecules free to interact with the kaolinite [6, 7].

The intercalation occurs in such a way that the molecules of the intercalating agent penetrate the kaolinite layers from the outside to the inside of the lamellar space. Therefore, very small kaolinite particles possess internal stresses (structural defects) in the crystallite that hinder the penetration of the host molecule, thus reducing the intercalation rate [8-11].

Other researchers have intercalated DMSO in kaolinite in order to obtain information about its behavior and structure. However, these studies have involved very long intercalation times, a fact that could, in principle, indicate difficulties in the use of kaolinite as a matrix [12-15].

In this work, the synthesization process of the nanocomposite precursor (K-DMSO) was modified, reducing the intercalation time to just 5 hours.

EXPERIMENTAL

The two-dimensional kaolinite matrix used here was treated kaolin (Premium) supplied by the company Caulim da Amazônia S.A. (CADAM), which was previously subjected to various stages of beneficiation, including dispersion and desanding, centrifugation, magnetic separation, chemical bleaching and flocculation, filtering, redispersion, evaporation and drying.

K-DMSO NANOCOMPOSITE SYNTHESIS METHODOLOGY

In the first procedure, 30 g of kaolin was placed in a round-bottomed 2-liter flask, to which a DMSO (9% v/v) solution was then added. The intercalation was carried out at room temperature under constant agitation.

The DMSO:H_2O ratio was kept constant at 9% v/v, since this value produced the best results in terms of the DMSO intercalation rate in the kaolin, according to the findings reported by Olejnik et al. [6].

Another experiment was carried out simultaneously under similar conditions, but the mixture was kept under heating at 95-100°C and agitation.

In order to evaluate the intercalation of DMSO molecules in the kaolinite lamellae, XRD and FTIR analyses were made of aliquots (15 ml) removed from the reaction at periodic intervals.

The samples were centrifuged at 3000 rpm for 20 min and oven-dried at 40-50°C for 4 days, ensuring that all the DMSO adsorbed on the kaolinite surface was removed, leaving behind only the intercalated molecules [16].

CHARACTERIZATION OF THE KAOLINITE AND NANOCOMPOSITES

Initially, the kaolinite was pressed on a glass slide, revealing its *(001)* plane. After the intercalation of the clay mineral with DMSO, the intercalated nanocomposites (K-DMSO) were again subjected to XRD in order to evaluate the variation of the basal interplanar distance and the Intercalation Rate.

The XRD analysis enabled us to determine an important parameter for evaluating the intercalation – the intercalation rate (II) –, which is calculated by the following expression:

$$II = [(In_1)/ (In_1 + K_1)] \times 100 \qquad (1)$$

where In_1 is the first basal reflection of the intercalated composite, and K_1 is the intensity of the first basal reflection in the non-intercalated matrix (original kaolinite).

The higher the value of II the more efficient the intercalation process, i.e., the greater the number of molecules intercalated between the lamellae of the kaolinite matrix. The thermal analyses were carried out with a Shimadzu TGA-50 thermogravimeter.

The kaolin and the nanocomposites were characterized by FTIR, using a Nicolet Magna IR 760 spectrometer. Solid samples (1 mg) were mixed with about 100 mg of dry pulverized potassium bromide (KBr) in an agate mortar. The resulting mixtures were pressed into pellets and used to obtain the spectra (% transmittance versus wave number) with a 4 cm^{-1} resolution and a scan rate of 32 kHz.

The kaolinite and nanocomposites obtained here were morphologically characterized by scanning electron microscopy. This technique is useful for observing the deformations in kaolinite crystallites that occur during the process of intercalation with dimethylsulfoxide and polymers. The chemical elements in the nanomaterial were identified by energy dispersive X-ray spectroscopy. The equipment used was a Zeiss DSM 940 digital scanning microscope and a JEOL NORAN equipped with EDS.

RESULTS AND DISCUSSION

1) INTERCALATION AT ROOM TEMPERATURE
a) X-RAY DIFFRACTION

Figure 3 shows the X-ray diffractograms obtained at 2θ, varying from 5 to 15°, which depict the basal interplanar distance of the kaolinite after intercalation with DMSO.

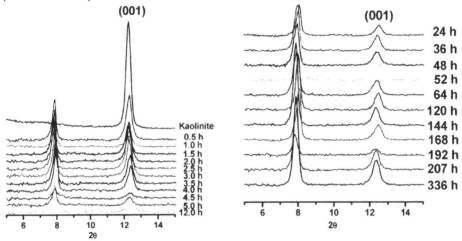

Figure 3: X-ray diffractograms of pure kaolinite and intercalated kaolinite nanocomposites (obtained without heating and at various intervals of time).

An analysis of these images reveals a shift of the peak corresponding to the (001) plane of the kaolinite, which was originally located at $2\theta = 12.45^{\circ}$ ($d = 0.72$ nm) and, after the intercalation, shifted to lower angles of $2\theta = 8.05^{\circ}$, i.e., $d = 1.11$ nm. Therefore, after the insertion of the DMSO molecules into the kaolinite lamellae, the interlamellar spacing increased by approximately 0.39 nm.

If the intercalation were complete, i.e., if an intercalation rate (II) of 100% were achieved, the initial peak at $2\theta = 12.45^{\circ}$ would disappear completely. However, even when kaolinite with only a small number of structural defects (well crystallized) is used, this value is not reached, and the maximum value reported in the literature is approximately 90%.

The peak corresponding to the (001) plane of kaolinite at 12.45° does not disappear completely, even over the longest period of intercalation considered (336). This can be explained by the presence of defects in the crystalline structure of kaolinite, which hinder the insertion of DMSO molecules in its interlamellar spaces.

The values of intensity of d = 0.72 nm and d = 1.11 nm were used and Equation 1 was applied to determine the intercalation rate (II), allowing for the construction of the graphs depicted in Figure 4.

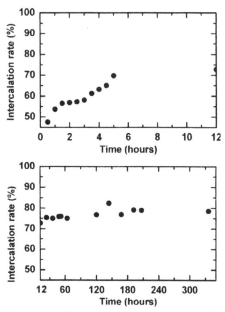

Figure 4: Variation of the intercalation rate as a function of time

In general, the intercalation rate increased over time, reaching a maximum value (82.6%) in 144 hours, i.e., in 6 days.

Gardolinski [17], who used kaolinite with a similar crystalline structure, obtained an intercalation rate of 83.5% but achieved this rate over a much longer time (480 hours).

We found that modifying the methodology usually employed in the intercalation of DMSO in kaolinite by introducing continued agitation throughout the entire process led to a significant reduction in the time required for the process (330% reduction in processing time).

The diffractograms corresponding to the best result achieved (144 hours) and the kaolin utilized are shown in Figure 5.

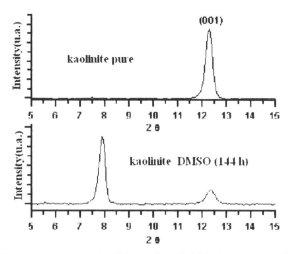

Figure 5: X-ray diffractograms of pure kaolinite and the K-DMSO nanocomposite obtained after 144 hours of intercalation without heating.

After the intercalation, we also noted an increase in background noise, which may indicate that the crystalline structure of the clay mineral particles underwent deformations (confirmed by SEM).

b)- FOURIER TRANSFORM INFRARED SPECTROSCOPY (FTIR)

Figure 6 shows the FTIR spectra for the pure kaolinite and the kaolinite/DMSO nanocomposites obtained at varying intercalation times. When intercalated in kaolinite, DMSO molecules interact with the hydroxyls of this clay mineral through hydrogen bridges. Hence, the characteristic intensity of the three bands corresponding to the hydroxyls (at 3694, 3667 and 3652 cm^{-1}) diminished.

In the K-DMSO nanocomposites, the band located at 3694 cm^{-1} decreased very little and the band at 3667 cm^{-1} (indicated by a red circle in Figure 6) disappeared. There was a slight reduction in intensity and broadening of the band at 3662 cm^{-1}, as well as the appearance of new bands at 3539 and 3502 cm^{-1} (dashed lines). The bands at 3020, 2935 and 2917 cm^{-1} correspond to stretching of the –C-H groupings of the DMSO molecules.

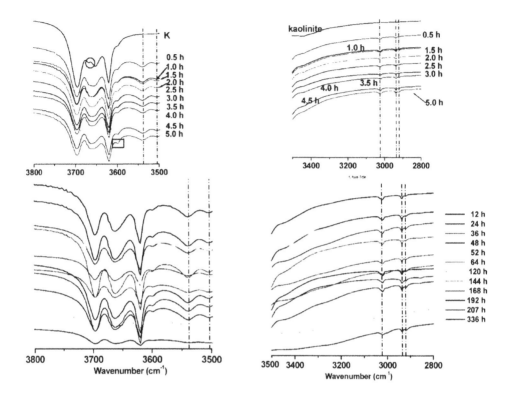

Figure 6: FTIR spectra of the pure kaolinite and the samples intercalated with DMSO (without heating)

c) SCANNING ELECTRON MICROSCOPY (SEM)

Figure 7 shows the SEM micrographs of kaolinite/DMSO obtained after 144 hours (without heating), magnified 10000, 20000 and 30000 times.

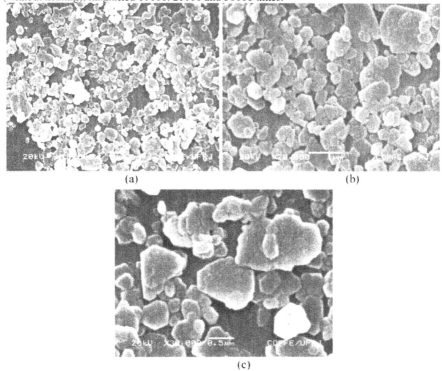

(a)   (b)

(c)

Figure 7: SEM photomicrographs of K- DMSO (144 hours) magnified:
(a) 10000 x, (b) 20000 x, and (c) 30000 x

After the intercalation process, a small number of particles retained their pseudo-hexagonal crystalline structure. The larger sized particles were found to present greater deformation (rounding of the edges of their crystallites), which was attributed to the stresses created by the insertion of the DMSO and to the intercalation process itself.

d) ENERGY DISPERSIVE X-RAY SPECTROSCOPY (EDS)

Figure 8 depicts the EDS results for the K-DMSO nanocomposite obtained after 144 hours of intercalation. Note the presence of Al, Si, O, Ti and Fe, C and S. The carbon came from the methyl groups, while the sulfur originated from the –S=) grouping present in the DMSO molecules.

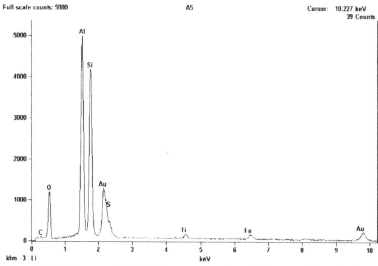

Figure 8: EDS image of K-DMSO nanocomposites (144 hours, without heating)

II) INTERCALATION AT 95-100°C

a) X-RAY DIFFRACTION (XRD)

Figure 9 shows the diffractograms obtained after the intercalation with DMSO under heating (95-100°C) at continual agitation.

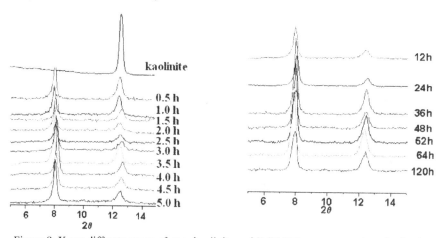

Figure 9: X-ray diffractograms of pure kaolinite and K-DMSO nanocomposites obtained under heating over various intercalation times

Note the shift of the peak corresponding to the (001) plane of the kaolinite, similar to the shift that occurred in the experiment without heating.

Figure 10 shows the variation of the intercalation rate as a function of time.

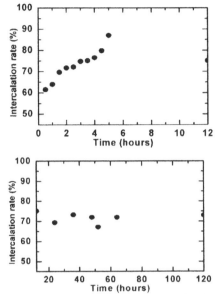

Figure 10: Variation of the intercalation rate as a function of time for the experiment carried out under heating and agitation

This figure shows a significant increase in the intercalation rate (87%) attained in just 5 hours of synthesis, i.e., a considerably more practical length of time than that obtained for the experiment at room temperature (82.6% in 144 hours of reaction).

Considering the values reported in the literature, this reduction was even greater, i.e., from 480 hours to 5 hours. In other words, a reduction of 9600% in the time required to obtain the K-DMSO precursor nanocomposite, with the advantage of attaining a high intercalation rate very close to that of well crystallized kaolinites.

Figure 11 shows details of the diffractogram of the pure kaolinite and the nanocomposite obtained in 5 hours of intercalation.

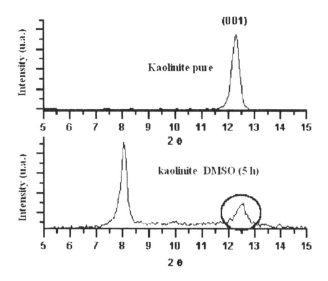

Figure 11: X-ray diffractogram of pure kaolinite and K-DMSO nanocomposite obtained
after 5 hours of intercalation (95-100°C, with agitation)

The figure clearly shows the increase in background noise after intercalation with
DMSO. A comparison of this diffractogram with that obtained in the experiment without
heating (Figure 4), both conducted under the best synthesization conditions (5 h and 144 h,
respectively), indicates that there was a greater increase in background noise for the
nanocomposite synthesized under heating. It can therefore be concluded that intercalation
under the effect of temperature leads to higher deformations in the crystalline structures.

Another piece of information that can be deduced by comparing these figures (Figures
4 and 11) has to do with the remanent non-intercalated kaolinite particles. Note that the peak
corresponding to the (001) plane of the remanent kaolinite (circled in blue in Figure 11)
underwent broadening, which can be considered an indication that temperature causes greater
deformation of its crystalline structure in a relatively short time (5 hours).

b) FOURIER TRANSFORM INFRARED SPECTROSCOPY (FTIR)
Figure 12 shows the FTIR spectra of the pure kaolinite and the K-DMSO
nanocomposites produced under heating, considering, for purpose of analysis, only intervals
of time of less than 5 hours due to the results obtained previously by XRD.

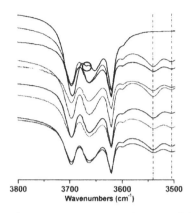 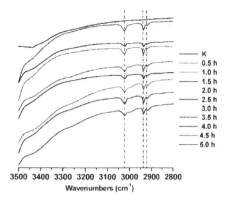

Figure 12: FTIR spectra of pure kaolinite and K-DMSO (with heating in different periods of time)

As occurred in the experiment without heating, there was a reduction in the intensity of the three surface hydroxyls bands corresponding to the wave numbers 3694, 3667and 3652cm^{-1}. and disappearance of the band at 3667 cm^{-1} (red circle). The band corresponding to the stretching of the internal hydroxyls (3619 cm^{-1}) remained unaltered. New bands were formed at 3539 and 3502 cm^{-1} (dashed lines), corresponding to the hydrogen bridges between the oxygen of the DMSO and the hydroxyls.

The bands at 3020, 2935 and 2917 cm^{-1} correspond to the stretching of the –C-H groupings of the DMSO molecules.

c) SCANNING ELECTRON MICROSCOPY (SEM)

The SEM image in Figure 13 depicts the K-DMSO nanocomposite obtained under heating in 5 hours of intercalation, magnified 30000 times.

Figure 13: SEM micrograph of K-DMSO obtained in 5 hours under heating (30000 x magnification)

After intercalation, the particles exhibited considerable deformation, losing their pseudo-hexagonal structure with blurred vertices, which is fully consistent with the X-ray diffraction results (Figure 11).

d) THERMOGRAVIMETRIC ANALYSIS (TGA)

Figure 14 shows the thermograms of kaolin samples, the DMSO:H₂O solution used in the intercalation, and the K-DMSO nanocomposite obtained under the best experimental condition, i.e., under heating at 95-100°C and continual agitation.

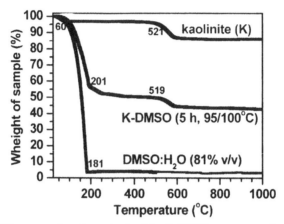

Figure 14: Thermogravimetric analysis of the K-DMSO nanocomposite compared with the original kaolin and the DMSO:H₂O solution

Although the reduction in the temperature of kaolinite dehydroxylation in the nanocomposite is an indication of a slight increase in the degree of structural disorder of kaolin when subjected to intercalation [18-20], the reduction shown in Figure 14 is very slight and, taken alone, would not be conclusive. However, the X-ray diffraction results (Figure 11) and the SEM photomicrographs (Figure 13) confirm the distortion of the kaolinite crystallites.

K-DMSO nanocomposites can be used as precursors in the synthesis of a vast range of nanomaterials using kaolinite as a two-dimensional matrix through displacement of the DMSO host molecule by the polymer to be intercalated. The principal objective of pre-intercalation with DMSO is to increase the interlamellar space of kaolinite, thereby facilitating the subsequent insertion of larger molecules.

Based on a critical review of the literature, we found that the K-DMSO nanocomposite, precursor of a series of other composites, is only obtained after long intercalation times, which generally vary from a few days to several months [5 – 7]. In most cases, these experiments were conducted at room temperature without agitation.

In the present work, the two routes employed led to the formation of K-DMSO nanocomposites in a shorter reaction time and produced a higher intercalation rate than the rates reported in the literature. When the synthesis is carried out under heating, the results are even more promising, overcoming the difficulties of using kaolinite as a matrix.

CONCLUSIONS

In this work, we evaluated the use of dimethylsulfoxide (DMSO) as an intercalating agent to obtain a kaolinite-DMSO (K-DMSO) nanocomposite precursor, and proposed that the best route involves heating (95-100°C) and continual agitation, which drastically reduces the time required for the intercalation to a mere 5 hours.

The X-ray diffraction results indicated that the intercalation rate achieved was 87%, which is close to the value obtained when using kaolinite with few structural defects.

The micrographs revealed that, after the intercalation process, the geometry and shape of the clay mineral crystallites were modified.

We concluded that it is possible to render kaolinite more attractive for the production of polymeric nanocomposites by means of modifications in the synthesization route of the K-DMSO precursor.

REFERENCES

[1] Almeida, A. M. L. F., *Modelização do Transporte de Carga em Polímeros Conjugados*, Tese de D. Sc., Universidade do Minho, Portugal, 2003.

[2] Rocha-Filho, R.C., *Química Nova na Escola*, n. 12, pp. 11-14, 2000.

[3] Pinho, M. S., Lima, R. C., Soares, B. G., Nunes, C. R., *Polímeros: Ciência e Tecnologia*, v. Out/Dez, pp. 23-26, 1999.

[4] Inzelt, G., Puskas, Z., *Elecrochemica Acta*, v. 49, pp. 1969-1980, 2004.

[5] Cases, F., Huerta, F., Garcés, P., Morallón, E., Vásquez, J. L., *Journal of Electrochemical Chemistry*, v. 501, pp. 186-192, 2001.

[6] Okubo, M., Fuji, S., Minami, H., *Colloid Polymer Science*, v. 279, n. 2, pp. 139-145, 2001.

[7] Cao, Y., Smith, P., Heeger, A. J., *Synthetic Metals*, v. 48, n. 1, pp. 91-97, 1992.

[8] Stejskal, J., Sapurina, I., *Pure Applied Chemistry*, v. 77, n. 5, pp. 815-826, 2005.

[9] Li, G., Pang, S., Xie, G., Wang, Z., Peng, H. Zhang, Z., *Polymer*, v. 47, pp. 1456-1459, 2006.

[10] Stejskal, J., Riede, A., Hlavatá, D., Prokes, D., Helmsteldt, M., Holler, P., *Synthetic Metals*, v. 96, pp. 55-61, 1998.

[11] Crispin, X., Cornil, A., Friedlein, R., Okudaira, K. K., Lemaur, V., Crispin, A., *Journal American Chemical Society*, v. 126, pp. 11889-11899, 2004.

[12] Stejskal, J., Saurina, I., Trchová, M., Konyushenko, E. N., Holler, P., *Polymer*, v. 47, pp. 8253-8262, 2006.

[13] Kokunov, Y. V., Gorbunova, Y. E., Khmelevskaya, L. V., *Russian Journal of Inorganic Chemistry*, v. 51, n. 12, pp. 1184-1190, 2005.

[14] Tagowska, M., Palys, B., Jackowska, K., *Synthetic Metals*, v. 142, pp. 223-229, 2004.

[15] Deng,Y, Whiteh, G. N., Dixon, J. B., *Journal of Colloid and Interface Science*, n. 250, pp. 379-393, 2002.

[16] Kocherginsky, N. M., Lei, W., Wang, Z., *Journal of Physical Chemistry A*, v. 109, pp. 4010-4016, 2005.

[17] Deng, Y, Dixon, J. B., White, N., *Journal of Colloid and Interface Science*, v. 257, pp. 208-227, 2003.

[18] Tzou, K., Gregory, R. V., *Synthetic Metals*, v. 47, n. 3, pp. 267-277, 1992.

[19] Gardolinsk, J. E. F .C, Carrera, L .C. M., *Journal of Materials Science*, v.35, pp.3113-3119, 2000.
[20] Gardolinsk, J .E. F. C, Ramos, L. P., Souza, G. P., Wypych, F., *Journal of Colloid and Interface Science*, v.221, pp. 284-290, 2000.

RAMAN SPECTROSCOPY OF ANATASE COATED CARBON NANOTUBES

Georgios Pyrgiotakis
Particle Engineering Research Center
University of Florida
Gainesville, FL, USA

Wolfgang M. Sigmund
Department of Materials Science and Engineering
University of Florida
Gainesville, FL, USA

ABSTRACT
    A high efficiency photocatalyst was synthesized by coating multiwall carbon nanotubes (CNTs) with particles of TiO_2 (anatase phase). Two distinct types of composites were synthesized with two different nanotube types as matrix: arc-discharge and chemical vapor deposition (CVD) grown CNTs. The particles were synthesized *via* sol-gel route and were then photocatalytically evaluated by the dye degradation technique. The particles with the arc-discharge core performed significantly better than the CVD CNTs (19.91 ± 0.20 min for the arc-discharge versus 177.41 ± 0.20 min for the CVD). The results for Raman spectroscopic characterization of the nanocomposite and its constituents are reported. Raman spectra show that both types of nanocomposite photocatalysts share chemical bonds of TiO_2 with the underlying CNT core. Furthermore, the spectra clearly indicate that the enhancement in photocatalytic activity was based on the difference in electronic properties of the CNTs. Arc-discharge synthesized CNTs were found to be metallic in nature whereas CVD grown CNTs were non-metallic.

INTRODUCTION
    In the last decade photocatalysis for environmental remediation has attracted a great deal of attention due to an increasing threat of chemical and biological hazardous constituents. They range from organic pollutants in the water to bacteria such as *E. Coli* or *B. Anthraces*. The current methodologies such as liquid disinfectants, filters and radiation, are very target oriented and cannot address all the risks. They rather focus on either remediation of chemical or on treatment of biological hazards. Photocatalysis has the advantage to treat both hazards simultaneously. In order to make photocatalysis work light and a photocatalyst is required. During the photocatalytic reaction a photon of proper energy ($h\nu \geq E_g$) strikes the surface of a semiconductor. Here, it generates an electron-hole pair (h^+-e^-) [1]. Nanocomposites are expected to take photocatalysis to the next level in environmental remediation overcoming current low efficiencies, which is especially true in treatment of biological samples [2, 3].
    Recent advances in photocatalytic enhancement in our group are based on engineered nanocomposites of titania and carbon. Multiwall carbon nanotubes (CNTs) were selected for their specific electronic properties that by design should impact the overall electronic state of the photocatalytic nanocomposite. Furthermore, CNTs act as electron carrier and as core material, serving as electron sink to remove the generated electrons and thus are expected to retard or completely prevent electron-hole recombination. This article focuses on the underlying electronic properties of a novel photocatalytic nanocomposite. Raman spectroscopy is introduced as main tool to characterize the key properties of the multiwall carbon nanotubes and their structural and electronic contributions to increased photocatalytic activity. The synthesis of carbon nanotube-titania nanocomposites was published earlier by us and other groups [4-6]. Although a brief summary of the experimental results and procedures will be given here, further references to the enhancement of photocatalytic efficiency

can be found for organic contaminants [7] (destruction of organic dye Procion Red MX-5B®) and for biological applications (destruction of *B.Cereus* spores) [8].

To demonstrate the hypothesis that the underlying electronic contributions from its constituents are key for photocatalytic enhancement, two electronically distinct types of carbon nanotubes were selected. The first one are pristine, arc-discharge synthesized multiwall carbon nanotubes (CNTs) while the second type is chemical vapor deposition (CVD) grown multiwall CNTs that have been further processed in a ball mill in the presence of strong mineral acids. This treatment significantly affects the properties of the tubes and further increases the differences in length. It especially shortens the CNTs. For short reference in the text from now on the arc-discharge CNTs will be referred to as long CNTs (ℓ-CNTs) while the ball milled CVD grown CNTs will be referred as short CNTs (*s*-CNTs). The photocatalytic efficiency of two types of nanocomposites consisting of an anatase coating on top of *s*-CNTs and ℓ-CNTs was evaluated by dye degradation tests.

The general theory of Raman spectroscopy accurately predicts the position of the Raman peaks of TiO_2. Titanium dioxide can exist in nine crystal structures [9], from which only anatase and rutile are reported to be photocatalytically active [2]. Each of these structures has very distinct vibrational frequencies. Anatase is tetragonal (D_{4h}^{19}) with two chemical formula units per unit cell and six Raman active modes ($A_{1g}+2B_{1g}+3E_g$). Rutile is also tetragonal (D_{4h}^{14}) and has a unit cell with four active modes ($A_{1g}+B_{1g}+B_{2g}+E_g$) [10]. Table I summarizes the peaks for anatase and rutile.

Table I: The Raman frequencies for anatase and rutile phase of titania. The notation in parenthesis represents the relative intensity of the peaks; w: weak; m: medium; s: strong; vs: very strong. [10]

Anatase D_{4h}^{19} 14_1/amd		Rutile D_{4h}^{14} P4$_2$/mnm	
E_g	144 cm^{-1} (vs)	B_{1g}	143 cm^{-1} (w)
E_g	197 cm^{-1} (w)	E_g	447 cm^{-1} (s)
B_{1g}	399 cm^{-1} (m)	A_{1g}	612 cm^{-1} (s)
A_{1g}	515 cm^{-1} (m)	B_{2g}	826 cm^{-1} (w)
B_{1g}	519 cm^{-1} (m)	-	-
E_g	639 cm^{-1} (m)	-	-

The Raman spectra of single and multiwall carbon nanotubes [11-13], both at the experimental [14-16] and theoretical [17] levels are well understood [12, 18, 19]. CNTs are one class of materials within just a few where experimental, theoretical and computational data are in complete agreement [19]. From the large number of peaks for the single wall carbon nanotubes the only peaks that remain in all multiwall CNTs are the G and D bands. The G band represents the 2-D features of graphite and appears also in CNTs and involves an optical phonon exchange between two dissimilar carbon neighboring atoms A and B in the unit cell [20-22]. The corresponding mode in the case of the tubular structure is the same. In contrast to the graphite structure, where the G band is a single frequency at around 1582 cm^{-1}, the nanotubes may show several peaks that relate to the relative position of the two carbon atoms on the tube. In general the frequencies that arise from vibration to a coaxial direction are lower compared to vibrations in circumferential direction [18]. The D band is one of the second order Raman scatterings and involves either one phonon (inelastic scattering) and one elastic scattering or two phonons [23]. The frequency where the Raman shift appears for the D band depends on the laser energy [24, 25]. A typical example of this feature is the D band at 1350 cm^{-1} that shifts by 53 cm^{-1}

when the laser energy changes by 1 eV [18]. For nanotubes this phenomenon appears at frequencies between 1305 cm^{-1} and 1350 cm^{-1} with full width at half maximum (FWHM) about 30-60 cm^{-1}.

2. MATERIALS AND METHODS

Synthesis of anatase coated CNTs

The arc-discharge CNTs were purchased from Alfa-Aesar (product #: 42886) in soot form while the CVD CNTs were obtained from Nanostructured & Amorphous Materials, Inc. (product #: 1236YJS) and were delivered in powder form. According to the manufacturer CVD CNTs were shortened via chemical mechanical shortening in a ball mill at the presence of nitric and sulfuric acid 1:3 ratio.

The anatase coatings on the CNTs were completed via sol-gel chemistry. The differences in the surface groups of the two distinct types of CNTs resulted into two slightly different chemical routes according to earlier publications [18]. For the coating of s-CNTs titanium iso-propoxide (Ti(OC$_3$H$_7$)$_4$) was used as precursor and titanium sulfate (Ti$_3$(SO$_4$)$_3$) in the case of the arc-discharge CNT. Both types of CNTs were treated in 60% (10 N) nitric acid (HNO$_3$) at 140°C to purify and chemically functionalize them prior to the coating application. The experimental protocol was as follows: 50 mg of the tubes were added to 200 ml of acid, which by the end of the process yielded approximately 30 mg of functionalized CNTs. The f-CNTs were kept under those conditions for approximately 10h at 140 °C, while the s-CNTs were held for 6h at 100 °C. The tubes were then washed three times with DI water (for the f-CNTs) or absolute ethanol (for the s-CNTs). The functionalization created surface groups such as −COOH, −OH and >C=O which are critical not only for the stabilization of the tubes in the solution, but also for the coating process since they can be used as initiation sites for the reactions. In addition they significantly enhanced the dispersion of the tubes in the medium where the coating took place.

The specific surface area of the CNTs was measured by BET. The titania precursor amount was based on the calculated total surface area of the CNTs. After the acid treatment the specific surface area for the f-CNTs and s-CNTs was 110 m^2/g and 135 m^2/g, respectively. The coating thickness was designed to be a 5 nm dense anatase layer on the CNTs, wherefrom the precursor amount was calculated. The pH of the reaction was selected based on the measured zeta potential of the tubes to achieve best achievable dispersion while maintaining a desirable reaction rate. Following that the f-CNTs were washed in DI water and after the last wash they were kept in suspension in a 300 ml three neck flask (30 mg of f-CNTs in 300 ml of DI water). The flask was placed in an oil bath at 40°C and was refluxed under constant stirring and the pH was adjusted at 3 using 0.1N HNO$_3$. Following, 106 µl of solution containing 45% wt Ti$_2$(SO$_4$)$_3$ in dilute H$_2$SO$_4$ (Sigma-Aldrich, product #:495182) were injected and the reaction was carried out for 1 hour. The s-CNTs were washed five times. The composite was then dried at 40 °C for two days. The process for s-CNTs was similar to the one followed for the f-CNTs. The solvent in this case, however, was ethanol since titanium isopropoxide reacts vigorously with water. In this case the pH was adjusted to approximately 4. In this case, 11.68 µl of water were added along with 58 µl of Ti(OC$_3$H$_7$)$_4$ (Sigma-Aldrich, product #:377996) and the reaction was carried out as before for 1 hr. After washing both samples were then heat-treated at 500°C for 3h with a ramping rate of 10 K/min. Detailed characterization and photocatalytic evaluation is available elsewhere [7].

Reference Material.

For Raman spectroscopy anatase nanoparticles (purchased from Sigma-Aldrich, product number: 44689) were used as reference material. The particles have an average Feret diameter of 5 nm and the size was chosen to be close to the coating thickness, since the 144 cm^{-1} line shape and location of the Raman spectrum for anatase is very sensitive to the size of the grain and therefore the size of the particles [26]. The size dependence is expressed by an asymmetric broadening of the peak line shape and a blue shift (towards higher wave numbers) [27]. Although in an infinite size crystal the phonons are free to travel in any direction before the lattice reabsorbs them, in the case of nano-sized crystals the phonons are confined in a space less than the minimum required for unconstrained interactions [28]. In reciprocal space the Raman line shape is given by the equation [27]:

$$I(w) = \int_{B.Z.} \frac{|C(0,q)|^2 \, d^3q}{[\omega - \omega(q)]^2 + a_L^2}$$

where B.Z. denotes the limits for the 1st Brillouin zone. a_L is the half width at half maximum, $\omega(q)$ is the phonon dispersion curve. $C(0,q)$ is the scattering coefficient for first order scattering and for spherical nanocrystals it can be written as:

$$|C(0,q)|^2 = \exp -\frac{q^2 d^2}{16 p^2}$$

This result is too complex to be directly calculated, but it can be approached with the assumption that the dispersive relation is a simple vibrational mode in a crystal, such as:

$$\omega(q) = \omega_0 + \Delta \times \left[1 - \cos\left(q \times a\right)\right]$$

Detailed calculations for particles of 5 nm average diameter predict a shift of 3.2 cm^{-1} towards higher values compared to the bulk titania, and an addition broadening of 3 cm^{-1} (FWHM) towards lower energy values [28]. This broadening and shift is related only to the 144 cm^{-1} line. Other bands are not affected by size.

EXPERIMENTAL PROCEDURES

Raman spectroscopy
   In every case 5 mg of photocatalyst were mixed with 1 ml of iso-propanol to form a slurry. The slurry was placed on a glass slide and left at room temperature to evaporate the iso-propanol. The Raman spectra were obtained using an InVia Renishaw microscope with laser wavelength of 785 nm. Since the samples were black in color the full power (65 mW) of the laser was used to maximize the acquired signal. Different spots of the same sample and different samples of the same material yielded similar spectra, with variations in intensities and noise levels. However, since none of the obtained peaks could be assumed as a fixed parameter the obtained spectra could not be averaged nor normalized.
   The measured spectra were smoothed using the LOWESS algorithm (locally weighted linear regression) with a quadratic polynomial and weighted least squares [29-31]. The Raman peaks were approached with Lorentz line shapes and in the case of the carbon nanotube and the G band the peaks were also fitted with the Breit-Wigner-Fano lineshape that is suitable for metallic nanotubes [32, 33]. In addition other lineshapes were used, but the results had low accuracy and further the semi-empirical nature had no significant meaning for the specific case. The baseline was modeled using a polynomial approach. Therefore the fitting equation was:

$$f(\omega) = \frac{1}{\pi} \sum_{i=1}^{n-1} \frac{I_0^{(i)} \alpha_0^{(i)}}{\left(\omega - \omega_0^{(i)}\right)^2 - \left(\alpha_0^{(i)}\right)^2} + \sum_{i=i}^{k} a_i \omega^i + I_0 \frac{1 + \dfrac{\omega - \omega_0}{q a_L}^2}{1 + \dfrac{\omega - \omega_0}{a_L}^2}$$

where I_0 is the intensity, $a_{L,0}$ is the half width at half height, ω_0 the frequency where the peak appears and q is a broadening parameter. The parameters i, the number of peaks, and j, background polynomial order, were manually selected. The regression parameters were $I0^{(i)}$, $a_{L,0}^{(i)}$, $\omega_0^{(i)}$, a_i and q. First, second and third order polynomials were tried to approach the background. The selection criterion was a minimum χ^2 value. Monte-Carlo Levenberg-Marquardt and Robust algorithms were used to fit the regression. In most of the cases all the algorithms gave the same fitting parameters with minor deviations. In some cases, certain algorithms (Monte-Carlo or Levenberg-Marquardt) failed to converge and only the remaining algorithms were used. The fit was attempted with more peaks besides the apparent number of peaks to ensure proper peak deconvolution.

RESULTS AND DISCUSSION
    As stated earlier, all samples were photocatalytically evaluated and were characterized via transmission electron microscopy (TEM), scanning electron microscopy (SEM), x-ray diffraction (XRD) and x-ray photoelectron spectroscopy (XPS) and some data was previously published [7]. Here we will present evidence to demonstrate CNTs important role in nanocomposite semiconductor photocatalysis. Therefore, focus is given to TEM and Raman spectroscopy and their analyses.

TEM analysis
    Figure 1 (a) and (b) show the TEM images of the CNTs after treatment with acids and before the coating step. The ℓ-CNTs appear to be straight and pristine with well defined walls while the s-CNTs are not straight, have a bamboo-like structure and the walls are curved and occasionally discontinued as consequence of the acid and ball mill treatment.
    The nanotubes' morphology and structure are features likely to affect their properties and in following the properties of the nanocomposite material. TEM analysis confirmed the manufacturer specifications that were provided, diameters of 20 nm for the s-CNTs and 15 nm for the ℓ-CNTs.
    Figure 1 (c) and (d) depicts both types of CNTs after the coating process. The same features underneath the anatase coating can still be viewed (the ℓ-CNTs are straight and well defined while the s-CNTs are curved and have damaged walls). However, there is a notable difference both with the coating thickness and its morphology. The coating appears thicker and rougher for the s-CNTs. This result was verified later with BET specific surface area measurements that indicated slightly larger specific surface areas for the anatase coated s-CNTs (183 m^2/g versus 162 m^2/g for the anatase coated ℓ-CNTs). The coating thickness was found to vary from 2 to 6 nm for the ℓ-CNTs while for the s-CNTs the thickness varied from 5 to 10 nm.

Photocatalytic activity.
    Detailed results for the photocatalytic activity of these nanocomposites have been reported elsewhere [8]. Here we report a summary of the data to demonstrate the significance of the impact. The interested reader is referred to Pyrgiotakis et al. [7] for the experimental protocol. Since photocatalysis is a surface chemical reaction the photocatalytic experiments were based on the same surface area. This experimental design allows investigating the impact of the electronic properties on the photocatalytic activity. The data in figure 2 were achieved by using the photocatalytic destruction

of a UV-stable dye with the commercial name of Procion Red MX-5B®. 1 mg of the coated *l*-CNTs (respectively 1.3 of the coated *s*-CNT and 3 mg of Degussa P25), was dispersed via sonication in 50 ml solution of 5 ppm dye and the suspension was placed in dark champer under the UV (375 nm). The dye concentration was measured indirectly by quantitative UV-VIS spectrophotometry. P25 (Degussa, Germany) was used as benchmark since it is widely accepted as an efficient photocatalyst. For data analysis it was assumed that the concentration reduction follows a simple exponential decay (Langmuir-Hinshelwood model).

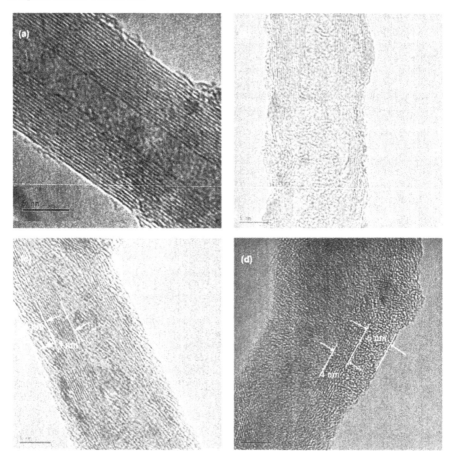

Figure 1: CNTs prior coating and with anatase coating. (a) Arc-discharge nanotubes (f-CNTs) (b) CVD grown nanotubes (s-CNTs) after the functionalization. (c) f-CNTs with anatase coating and (d) s-CNTs with anatase coating.

The experiments were performed under 375 nm UV light with total intensity of 20 W/m². For the Degussa P25 in addition to the 375 nm wavelength, the 350 nm lamp, since it is the most commonly used wavelength. The results are plotted as decrease in concentration of the dye with time of UV exposure. Continuous lines are fits to the Langmuir-Hinshelwood model ($C=C_0 \, Exp(-t/\tau)$). The data can be best compared as τ, which is the time required for a 63.21% (1/e) reduction of the dye concentration.

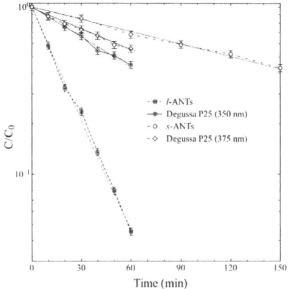

Figure 2: Collective representation of the photocatalytic dye degradation results using anatase coated CNTs and P25 titania.

The coated ʃ-ANTs yield a τ of 19.91±0.40 min (reaction rate 4×10^{-8} mols/min·mg) while the s-CNTs had a τ of 177.41±10.00 min (reaction rate 4.63×10^{-9} mols/min·mg). Since the measurement was done on the same surface area, it implies that the difference between the two CNT nanocomposite particles (789% decrease in the efficiency) is caused by the different types of nanotubes. The results for the P25 particles were 24.1±0.4 min under the 350 nm (reaction rate 3.4×10^{-8} mols/min·mg) and 97.0±2.7 min (reaction rate 8.4×10^{-9} mols/min·mg) falling in between the CNT nanocomposites. To explain these dramatic differences in photocatalytic performance Raman spectroscopic analysis of the nanocomposites was applied.

Raman spectra
Figure 3 shows the Raman spectra for the carbon nanotubes segment (1000-1800 cm⁻¹) prior ((a), (b)) and after coating ((c), (d)), were (a) and (c) are for the ʃ-CNTs and (b) and (d) are for the s-CNTs. As previously stated the coating for the s-CNTs was thicker so the obtained Raman spectrum was not as intense as for the ʃ-CNTs therefore the signal to noise ratio is significantly lower. In all cases, however, both G and D bands were observed. The data presented are the results after applying smoothing algorithms with the parameters that influence the shape and the position of the peaks the

least. For the uncoated l-CNTs the D band appeared at the 1312 cm^{-1} with an a_L of 22 cm^{-1} which is a very good indication that the tubes were pristine. In addition the G band at 1594 cm^{-1} had a very distinct split (G$^-$ at 1583 cm^{-1} and G$^+$ at 1611 cm^{-1}). Breit-Wigner-Fano fitting gave better results for the G$^-$, which is characteristic for the metallic nature of the carbon nanotubes. This is one of the most important results since, it points out that the l-CNTs are electrically conductive.

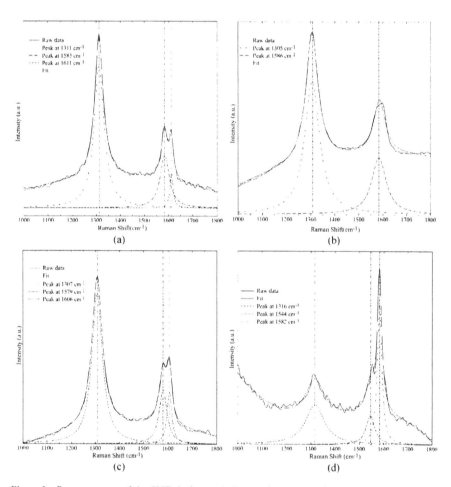

Figure 3: Raman spectra of the CNTs before and after coating. (a) arc-discharge nanotubes (l-CNTs) (b) CVD grown nanotubes (s-CNTs) after ball milling in mineral acid. (c) l-CNTs with anatase coating and (d) s-CNTs with anatase coating.

For the uncoated s-CNTs the D band appears at 1305 cm⁻¹, which is at the upper limit of the span were the D band could appear in CNTs. The regression gave an a_1 value of 31 cm⁻¹, which is broader than what was expected for CNTs but it is reasonable due to the damage of the ball mill treatment. The next characteristic is the G-band that appears at 1586 cm⁻¹ with an a_L of 30 cm⁻¹. Although from the profile of the peak it can be hypothesized that there were two peaks, all the fit algorithms failed to recognize two peaks with variation in the smoothing parameters and background. It is, therefore, accurate to conclude that there is not a distinct split of the background. After the coating the l-CNTs had not changed significantly with only some shifting of the peaks and differences of the relative peak intensities. Similarly the results from the s-CNTs also displayed differences in the peak relative intensities and the peak location after the anatase coating. However it should be noted here that the second peak that appears for the case of the s-CNTs is seems to be an artifact, since it is not justified by the prior coating spectrum. So the G+ is the original G-Band, which has shifted from the original position by 4 cm⁻¹ due to the coating and the possible bond between the coating and the CNT.

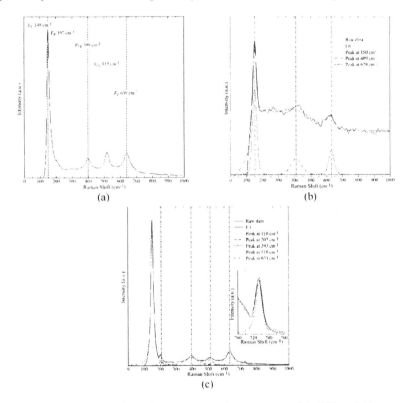

Figure 4: Raman spectra of (a) reference anatase, (b) anatase coated l-CNTs and (c) anatase coated s-CNTs. The insert shows an enlargement of the 700-760 cm⁻¹ region.

Figure 4 shows the Raman spectra obtained by the reference sample and the titania segment (0-1000 cm^{-1}) of the coated tubes. All the peaks for the reference material appear in the locations listed in Table I with only one exception where the 144 cm^{-1} peak has been shifted to 148 cm^{-1}. Considering that the manufacturer gives particle size 5 nm, this is in agreement with the theoretical prediction by Choi et al [34]. Since the coating process of the s-CNTs yielded more coating than the f-CNTs the titania peaks appeared stronger and sharper while for the f-CNTs only the very strong peaks were observed. Again all titania peaks seem shifted compared to the reference and the shifting is either red or blue. In addition, a new peak appeared for the s-CNTs at 727 cm^{-1}, which could not be identified as any known peak for anatase, rutile, or CNTs (see insert in figure 4(c)).

Based on the TEM images and Raman spectra it can be concluded that the f-CNTs in this work are electrically conducting since the split at the G band is very distinct. They also have a well defined tubular structure which fits the G and D band location and width. Finally they are characterized by very high density of surface defects as a result of the acid treatment as is demonstrated by the relative peak intensities. On the contrary, the apparent absence of the split (or at least a very distinct split) for the s-CNTs underlines their poor electrically conductivity.

The second important result is the peak shifting that occurred after coating for both types of CNTs. The peaks have different shift not only in magnitude, but in direction, too. This not only eliminates the possibility of instrumental error, but also shows the existence of many complicated bonds. In more detail the G^{+} band arises from the carbon atoms vibrations along the nanotube axis, and the frequency is sensitive to charge transfer from dopant addition with up-shift for acceptors and downshift for donors [40]. In the case of the f-CNTs the G^{+} band shifts from 1612 cm^{-1} to 1606 cm^{-1}. This might be explained by titania generated electrons, triggered by the ambient light or the laser, that can be trapped in the CNT structure and may cause an energy up-shift of the G^{+} band.

In both cases the E$_g$ titania peak appeared at 150 cm^{-1}, which indicates a 6 cm^{-1} blue shift compared to the literature value and a 2 cm^{-1} shift compared to the reference material. The possible reasons for this shift are the size constraint and possible chemical interactions. The surface termination of titania particles imposes constraints to the phonons, which results in a more asymmetric peak and blue shift. However, a shift of this magnitude has been calculated according to the theory described previously and is in the order of 3 nm. Consequently it cannot be assumed that the blue shift comes only from the size restrictions. In addition all the titania peaks for both materials are significantly shifted toward either direction and with different values for each peak. Therefore, the Raman spectra are clearly showing TiO$_2$/MWNTs interactions. The most likely interaction is a bond such as C−O−Ti. Yakovlev et al. [34] mention the Ti-O-Si bond at 950 cm^{-1}. Similarly here the 727 cm^{-1} may be attributed to a Ti−O−C bond. This is consequence of the method followed for the coating, initiated by the −OH and >C=O formation on the CNT surface. This bond is required for C to act as dopant to titania by making titania more effective by means of reducing the minimum required photon energy to generate the electron hole pair [35].

CONCLUSIONS

To increase the photocatalytic efficiency nanocomposites of TiO$_2$ and multi-wall carbon nanotubes (CNTs) were synthesized. The CNTs are the support structure for titania nanoparticles (anatase phase) and act as a sink for the generated electrons. The goal was to use the high surface area of the nanotubes and their metallic properties to increase the photocatalytic efficiency. The metallic nature of the tubes makes them an excellent scavenger of the electrons generated during the UV irradiation and therefore extends the lifetime of the holes. To investigate the impact of the nanotube nature in the current approach we used both arc discharge and CVD nanotubes. The arc discharge CNTs are known to have excellent metallic properties, which is an immediate result of the high synthesis temperature, while the CVD tubes in general are nonmetallic. This is demonstrated by a distinct split of the G-band in the Raman spectrum for the arc-discharge CNTs which is a significant characteristic for metallic carbon nanotubes.

The Raman spectrum after anatase coating showed a shift of all the major peaks of titania and carbon nanotubes and one extra peak at 727 cm^{-1}. All these characteristics can be attributed to a Ti–O–C bond. This bond is very significant not only because it acts as a channel to allow the electrons to be transported from the titania to the carbon nanotube core, but allows the carbon to perform as dopant to the titania. In general, the synthesized nanocomposite acts as an enhanced photocatalyst in a multitude; the specific surface area increases due to the needle like shape of the CNT support; the nanotube acts as an electron sink and scavenges the electrons.

ACKNOWLEDGEMENTS

The authors would like to acknowledge support by the Particle Engineering Research Center (PERC) of the University of Florida and the National Science Foundation under Grant No. EEC-94-02989. The authors would like to acknowledge Kerry Siebein from MAIC for the TEM images.

REFERENCES

1. Hoffmann, M.R., et al., Environmental Applications of Semiconductor Photocatalysis. *Chemical Review*, **95**(1): p. 69-96 (1995).
2. Fujishima, A., T.N. Rao, and D.A. Tryk, Titanium dioxide photocatalysis. *Journal of Photochemistry and Photobiology C*, **1**: p. 1-21 (2000).
3. Linsebigler, A.L., G. Lu, and J.J.T. Yates, Photocatalysis on TiO$_2$ surfaces: principles, mechanisms, and selected results. *Chemical Review*, **95**(3): p. 735-758 (1995).
4. Jitianu, A., et al., Synthesis and characterization of carbon nanotubes - TiO2 nanocomposites. *Carbon*, **42**(5-6): p. 1147-1151 (2004).
5. Jitianu, A., et al., New carbon multiwall nanotubes - TiO$_2$ nanocomposites obtained by the sol-gel method. *Journal of Non-Crystalline Solids*, **345-46**: p. 596-600 (2004).
6. Lee, S.W. and W.M. Sigmund, Formation of anatase TiO$_2$ nanoparticles on carbon nanotubes. *Chem Commun (Camb)*, (6): p. 780-1 (2003).
7. Pyrgiotakis, G., S.H. Lee, and W.M. Sigmund, *Advanced Photocatalysis with Anatase Nanocoated Multi-walled Carbon Nanotubes*, in *MRS Spring Meeting 2005*, Lu S., et al., Editors. 2005: San Fransisco.
8. Lee, S.H., et al., Inactivation of bacterial endospores by photocatalytic nanocomposites. *Colloids and Surfaces B-Biointerfaces*, **40**(2): p. 93-98 (2005).
9. Muscat, J., V. Swamy, and N.M. Harrison, First-principles calculations of the phase stability of TiO$_2$. *Physical Review B*, **65**(22): p. 224112-1 (2002).
10. Chaves, A., R.S. Katiyar, and S.P.S. Porto, Coupled Modes with A1 Symmetry in Tetragonal BaTiO$_3$. *Physical Review B*, **10**(8): p. 3522-3533 (1974).
11. Belin, T. and R. Epron, Characterization methods of carbon nanotubes: A Review. *Materials Science and Engineering B-Solid State Materials for Advanced Technology*, **119**(2): p. 105-118 (2005).
12. Dresselhaus, M.S., et al., Raman spectroscopy on isolated single wall carbon nanotubes. *Carbon*, **40**(12): p. 2043-2061 (2002).
13. Dresselhaus, M.S., et al., Raman spectroscopy of carbon nanotubes *Physics Reports*, **409**: p. 47-99 (2005).
14. Corio, P., et al., Spectro-electrochemical studies of single wall carbon nanotubes films. *Chemical Physics Letters*, **392**(4-6): p. 396-402 (2004).
15. Dresselhaus, M.S., et al., Single nanotube Raman spectroscopy. *Accounts of Chemical Research*, **35**(12): p. 1070-1078 (2002).
16. Dresselhaus, M.S., et al., Science and applications of single-nanotube Raman spectroscopy. *Journal of Nanoscience and Nanotechnology*, **3**(1-2): p. 19-37 (2003).

17. Saito, R., G. Dresselhaus, and M.S. Dresselhaus, *Physical Properties of Carbon Nanotubes*. 1998, London: Imperial College Press.
18. Dresselhaus, M.S., et al., Raman spectroscopy of carbon nanotubes. *Physics Reports-Review Section of Physics Letters*, **409**(2): p. 47-99 (2005).
19. Knight, D.S. and W.B. White, Characterization of Diamond Films by Raman-Spectroscopy. *Journal of Materials Research*, **4**(2): p. 385-393 (1989).
20. Dresselhaus, M.S. and G. Dresselhaus, Intercalation Compounds of Graphite. *Advances in Physics*, **30**(2): p. 139-326 (1981).
21. Enoki, E., M. Endo, and M. Suzuki, *Graphite Intercalation Compounds and Applications*. 2003, New York: Oxford University Press.
22. Jorio, A., et al., Resonance Raman spectra of carbon nanotubes by cross-polarized light. *Physical Review Letters*, **90**(10): p. 107403.1-107403.4 (2003).
23. Dresselhaus, M.S. and P.C. Eklund, Phonons in carbon nanotubes. *Advances in Physics*, **49**(6): p. 705-814 (2000).
24. Kürti, J., et al., Double resonant Raman phenomena enhanced by van Hove singularities in single-wall carbon nanotubes. *Physical Review B*, **65**(16): p. 165433.1-165433.9 (2002).
25. Milnera, M., et al., Periodic resonance excitation and intertube interaction from quasicontinuous distributed helicities in single-wall carbon nanotubes. *Physical Review Letters*, **84**(6): p. 1324-1327 (2000).
26. Zhang, W.F., et al., Raman scattering study on anatase TiO_2 nanocrystals. *Journal of Physics D-Applied Physics*, **33**(8): p. 912-916 (2000).
27. Bersani, D., P.P. Lottici, and X.Z. Ding, Phonon confinement effects in the Raman scattering by TiO2 nanocrystals. *Applied Physics Letters*, **72**(1): p. 73-75 (1998).
28. Choi, H.C., Y.M. Jung, and S.B. Kim, Size effects in the Raman spectra of TiO_2 nanoparticles. *Vibrational Spectroscopy*, **37**(1): p. 33-38 (2005).
29. Cleveland, W.S. *Visual and Computational Considerations in Smoothing Scatterplots by Locally Weighted Fitting*. 1978 [cited.
30. Cleveland, W.S., Robust Locally Weighted Fitting and Smoothing Scatterplots. *Journal of the American Statistical Association*, **74**: p. 829-836 (1979).
31. Cleveland, W.S., LOWESS: A Program for Smoothing Scatterplots by Robust Locally Weighted Fitting. *The American Statistician*, **35**: p. 54 (1981).
32. Brown, S.D.M., et al., Origin of the Breit-Wigner-Fano lineshape of the tangential G-Band feature of metallic carbon nanotubes. *Physical Review B*, **63**(15): p. 155414.1-155414.8 (2001).
33. Dubay, O. and G. Kresse, Accurate density functional calculations for the phonon dispersion relations of graphite layer and carbon nanotubes. *Physical Review B*, **69**(8): p. 035401 (2004).
34. Yakovlev, V.V., et al., Short-range order in ultrathin film titanium dioxide studied by Raman spectroscopy. *Applied Physics Letters*, **76**(9): p. 1107-1109 (2000).
35. Souza, A.G., et al., Raman spectroscopy for probing chemically/physically induced phenomena in carbon nanotubes. *Nanotechnology*, **14**(10): p. 1130-1139 (2003).

STRUCTURAL AND OPTICAL PROPERTIES OF SOL-GEL DERIVED HYDROXYAPATITE FILMS IN DIFFERENT STAGES OF CRYSTALLIZATION AND DENSIFICATION PROCESSES

Stoica Tionica[1]; Gartner, Mariuca[2]; Ianculescu, Adelina[3]; Anastasescu, Mihai[2]; Slav, Adrian[1]; Pasuk, Iuliana[1]; Stoica, Toma[1]; Zaharescu, Maria[2]

[1]National Institute of Materials Physics, Bucharest, Romania.
[2]Institute of Physical Chemistry "Ilie Murgulescu", Romanian Academy, Bucharest, Romania.
[3]Materials Science and Engineering, Politehnica University of Bucharest, Bucharest, Romania.

ABSTRACT

Hydroxyapatite (HAp) films, widely used as a biocompatible material for hard tissues repairing, are studied in this paper for a better control and understanding of the crystallization and densification of sol-gel deposited layers. Calcium nitrate and triethyl phosphite diluted in alcohols were used as calcium and phosphorus precursors. HAp coatings were obtained by spinning method, followed by drying and annealing in the range 130-750°C. Specific temperatures of the chemical reactions of HAp formation have been revealed by thermogravimetric analysis (TG-DTA). Structural, chemical and optical characterizations of the films were performed by Scanning Electron Microscopy (SEM) with Energy Dispersive X-ray Analysis (EDX) facilities, X-ray diffraction (XRD), Fourier Transform Infrared Spectroscopy (FTIR) and Spectral Ellipsometry (SE). The film crystallinity increases with the annealing temperature above 550°C. XRD lines of HAp crystallographic planes become narrow at 750°, and correspond to a single crystallographic phase. The film morphology depends not only on annealing history but also on the synthesis parameters. Thus, layers with large grains of well sinterized HAp nano-crystals can be obtained. Using EDX measurements, the Ca/P ratio was found close to the stoichiometric value in the well sinterized HAp grain regions and slightly increased 1.82 for boundary regions between grains. The density of pores usually observed in sol-gel derived films was estimated for different samples using modeling of SE data.

INTRODUCTION

Hydroxyapatite with the chemical formula $Ca_{10}(PO_4)_6(OH)_2$ is one of the most important biomaterial used for orthopedic and dental applications due to its chemical similarity with bones and its bioactive properties.[1-3] Plasma spray technique is currently used for HAp coatings. But this technique, due to a high-temperature process, has some disadvantages such as deviation from stoichiometry, crystalline phase and homogeneity.[4] Therefore, other techniques including electrodeposition,[5] pulsed laser deposition[6] and sputtering deposition[7] have been studied and reported in literature.

Recently, sol-gel method has been developed for synthesizing of hydroxyapatite.[8-11] By this method, a very good control of the composition and a lower temperature crystallization[9,12] of HAp films can be achieved in comparison with conventional methods such as hydrothermal synthesis and wet precipitation.[13] Using sol-gel method, hydroxyapatite coatings as well as hydoxyapatite composites can be obtained depending on the chemical parameters.[14-18]

The commercial titanium implants have good mechanical properties but poor osteointegrative properties. HAp coatings can solve this problem. Moreover, a TiO_2 buffer layer is a good barrier against metal ions diffusion in the human body, and offers a chemical and mechanical stability.[19,20]

In this paper, optimization of sol-gel process in order to obtain single phase HAp crystalline coatings with different porosities is studied. For this, useful information is obtained by studies of HAp powders obtained from gel. The film porosity was varied using different annealing parameters or/and Ti and TiO_2 buffer layers.

EXPERIMENTAL DETAILS

For sol-gel deposition, solutions containing 0.2M $Ca(NO_3) \cdot 4H_2O$ and triethyl phosphite $P(C_2H_5O)_3$ in basic medium was used. After spin coating (1000 rpm) of the sol on Si(100) substrates at room temperature, the wafers were dried for 15-30 min. at 150°C and annealed in air atmosphere for 30-60 min. at 550°C and higher temperatures. The solution parameters and annealing procedure were optimized for a complete removal of solvents and the formation of single crystalline phase HAp films. Thermo-analysis of HAp powders was performed using Pyris-Diamond derivatograph The crystalline structure was investigated by XRD measurements using a BRUKER-D8 Advance type X-ray instrument. The optical properties were determined by spectroellipsometric measurements in the spectral range 0.4-0.7µm. For FTIR measurements a Perkin Elmer Spectrum BX instrument was used. A Hitachi S2600N model equiped with Rontec Edwin WinTools was used for SEM and EDX studies.

TG-DTA AND XRD STUDIES OF SPECIFIC ANNEALING TEMPERATURES AND CRYSTALLIZATION OF SOL-GEL DERIVED HAp POWDER

After spinning deposition of thin gel layer on different substrates, an annealing process is necessary for removal of additional components and crystallization of HAp films. In order to find out specific reaction temperatures of the HAp formation using the sol-gel method, we have investigated by TG-DTA a powder obtained after an evaporation of the precursor solution. The same slightly basic (pH 7 – 8) solution used for HAp film deposition has been transformed in a powder before TG-DTA measurements. In Figure 1, the weight (TG) and differential thermo-analysis (DTA) during annealing process are shown as a function of temperature. Weight losses shown by the TG curve (Figure 1a) can be associated with exo- and endo- thermic reactions revealed by DTA curve (Figure 1b).

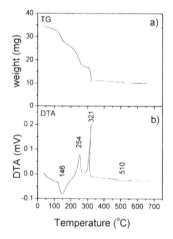

Figure 1 - Temperature dependence of the TG (a – weight) and DTA (b – differential thermo-analysis) signals during annealing of the precursor HAp powder (maximum annealing rate of 10 °C/min).

One can see from Figure 1a that, in spite of initial drying of the powder, a significant weight loss still exist within the temperature range $100 - 130\ ^0C$, due to elimination of water and organic solvents. Decomposition of nitrate compounds occurs within the temperature range $130 - 280\ ^0C$, while the combustion and elimination of organic compounds is obtained at about 321^0C.[15]

The powder weight continues to slightly decrease (~1%) within the high temperature range (350 – 750 ^{0}C), without a specific energy. However, an endo-thermic reaction at 510°C can be observed in DTA diagram. At 550°C, the HAp powder is already in a poli-crystalline form but its crystallinity increases constantly with the annealing temperature, as shown by narrowing of the XRD lines. SEM image of the HAp powder after annealing at 750°C is shown in Figure 2. The powder consists in large grains with sizes of a few micrometers, which are formed by a compact agglomeration of nanocrystals of about 100nm.

Figure 2 - SEM image of the HAp powder after annealing at 750 °C

Figure 3 shows a XRD curve of HAp powder, after annealing at 750°C. The diffraction lines correspond to a single crystalline HAp phase, as shown by solid dots as marks of different diffraction angles in agreement with standard data JCPDS 9-0432 for HAp powder.

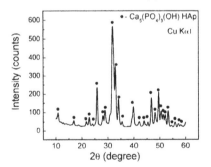

Figure 3 - XRD curves of the HAp powders after annealing at 750°C:

Studies performed on powder are very usefully not only for a design of the annealing procedure of the HAp film formation, but also to detect possible multiphase behaviors induced by deviation of different parameters of the precursor solution, from optimal values. For example, an incomplete hydrolysis may result in a multiphase composition of the powder (additional components calcite and pyrophosphate, data not shown here).

FTIR AND XRD STUDIES OF SOL-GEL DERIVED HAp FILMS
The annealing effect on sol-gel derived HAp films was studied using FTIR spectroscopy. Figure 4 shows the spectral absorption curves of sol-gel HAp films after annealing at different temperatures.

The curves are shifted on vertical axis. for clarity. After annealing at 150 °C. water and other components are still massively present within films, in agreement with TG-DTA studies on powders. The spectrum is essentially changed after annealing at 300 °C and has a smooth evolution with temperature above 300 °C. No other sharp transition is observed within the range 300 – 550 ^{0}C. in contrast with the results of the TG-DTA measurements which show an essential exothermic transformation at 321 °C. It might be that, due to a reduced thickness of the film. this transformation occurs already at lower temperature 300°C.

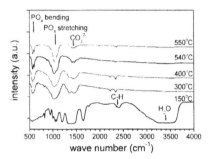

Figure 4 - FTIR spectra of sol-gel derived HAp films for different annealing temperatures

The annealing at higher temperatures results in a constant decreasing of the C-H absorption which is eliminated by annealing at 550 °C. The absorption peaks associated to bending and stretching oscillations of the PO_4 group become narrower at higher annealing temperatures. corresponding to more compact HAp films. Interesting, the absorption band associated to CO_3^{-2} group is observed in FTIR spectra, in spite of a single crystalline HAp phase detected by XRD measurements. after high temperature annealing. as we discuss downward. We have to admit that this group is included either as an impurity in the crystalline phase or in an amorphous matrix.[16]

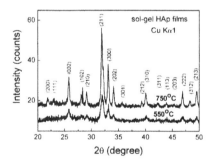

Figure 5 - XRD grazing incidence measurements on sol-gel HAp films after annealing at 550°C and 750°. Labels are crystalline planes associated to HAp diffraction lines.

XRD measurements at a grazing incidence of 2° are shown in Figure 5 for HAp films annealed at 550 °C and 750 °C. The crystallinity of the film is increased for higher annealing temperature. as can be deduced from the increase of the peak intensities and narrowing of the direction lines. The diffraction lines in Figure 5 are associated to different crystalline planes in agreement with standard PDF data files for HAp. One can see that the film is a single HAp crystalline phase, and no essential contribution of possible other components (calcite and pyrophosphate) can be detected.

By increasing the annealing temperature from 550°C to 750°C, not only the crystallinity of the film is increased, but also the morphology of the film is modified. In Figure 6, SEM images are comparatively shown for films annealed at 550°C to 750°C. The film is reorganized from a fine structured morphology (Figure 6a) into more extended compact areas by annealing at higher temperature (Figure 6b). However, some cracks are observed in films annealed at higher temperatures. These behaviors are specific for films obtained using slightly basic precursor solution (pH ~8.5).

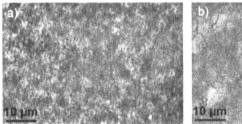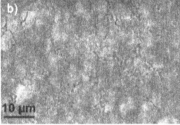

Figure 6 - SEM images of sol-ge HAp films using slightly basic precursor solution (pH ~8.5), after annealing at: a) 550°C; b) 750°C.

Using a precursor solution of less basic character (pH ~7.5), larger grains of compact HAp (diameter of 10 - 20 μm) can be obtained in films annealed at lower temperatures, as can be seen as dark areas in SEM images of a film annealed only at 550°C (Figure 7). The grains are interconnected by more porous boundary regions (bright areas in Figure 7).

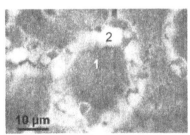

Figure 7 – SEM image of a HAp film after annealing at 550°C, using a reduced pH 7.5 of the precursor solution. Large compact regions (label 1) and more porous boundary areas (label2) are observed.

Compact and boundary regions (in Figure 7, labels 1 and 2, respectively) have been comparatively investigated by EDX. Within regions of more compact HAp, the value of the Ca/P ratio was found ~1.64, close to the stoichiometric value 1.66, while for boundaries, this ratio has a slightly increased value of about 1.82. Since only the crystalline structure of HAp phase was detected by XRD measurements we can conclude that the deviation from stoichiometric value of the boundary regions does not essentially influence the crystalline structure of the film.

ELLIPSOMETRY STUDIES OF OPTICAL AND COMPOSITION PROPERTIES OF HAp FILMS

Optical properties were studied using SE within the spectral range $0.4 - 0.7$ μm. The main aim of these studies was an estimation of composition of HAp films. For our case of single crystalline HAp films, a mixture of HAp and voids (air) was assumed. Computed ellipsometric functions using multicomponent Bruggemann Effective Medium Approximation (B-EMA)[21] of optical constants of each layer were fitted to the experimental data.

Table I – Thickness (d) and composition (HAp and voids) of HAp films evaluated by EMA fitting of ellipsometry data for different samples. HAp films were obtained by 1 or 2 succesive depositions (number of HAp layers) on c-Si substrates with or without a buffer layer. Different annealing regims were used (Ann-1 and Ann-2: annealing temperature ramp 7°C/min and 3°C/min, respectively).

Sample	Buffer layer	number of HAp layers	Comments	d (nm)	HAp (%)	voids (%)
HA-Si-1	-	1	Ann-1	187	40.4	59.77
HA-Si-2	-	1	Ann-2	95	70.6	29.4
HA-Si-3	-	1	Ann-2	108	67.2	32.8
HA-Si-4	-	2	first layer Ann-1; second layer Ann-2;	226	70.7	29.3
HA-Si-5	-	2	Ann-2	218	79.8	20.2
HA-Ti-1	Ti	1	Ann-2	348	64.3	35.7
HA-Ti-2	Ti	1	Ann-2	401	55.5	44.5
HA-Ti-3	Ti	2	Ann-2	746	61.7	38.3
HA-TiO2-1	TiO_2	1	Ann-2; buffer 39nm TiO_2 with 4.6% porosity	191	94.9	5.1
HA-TiO2-2	TiO_2	1	Ann-2; buffer 163nm TiO_2 with 0.7% porosity	232	99.6	0.4

Samples of HAp films deposited on different substrates and different annealing procedures, for 1 or 2 successive HAp sol-gel depositions have been investigated (Table I). Oxidizes c-Si wafers (native or thermal-grown oxides) with or without a buffer layer were used as substrates. Silicon wafers have been covered before HAp deposition, either by vacuum evaporation of Ti layers 50nm thick, or by sol-gel deposition of TiO_2 layers.

The substrate type, as well as the temperature rate of the annealing procedure strongly influences the porosity of HAp films, as can be seen in Table I. In Table I, beside some information about the sample preparation, ellipsometry fitting results on thickness and porosity (void concentration) are given.

The used fitting procedure takes into account all layers of the sample structure. For all investigated samples a weak dispersion of the refractive index of the HAp layer was found, corresponding to a large bandgap of the HAp, in agreement with literature data. However, the value of the refractive index is strongly influenced by both, the annealing procedure and the substrate type.

Dispersion curves of the refractive index are shown in Figure 8a for different samples described in Table I. The refractive index varies within the range $1.21 - 1.56$ from sample to sample. In Figure 8b the dispersion curves are shown on extended scale for samples with dense HAp films deposited on TiO₂ buffer.

The fluctuation of the refractive index is correlated with void concentration. For substrate without buffer, a reduction of the temperature rate of the annealing procedure results in a strong decrease of the layer porosity from 60% to less than 30% (see Table I).

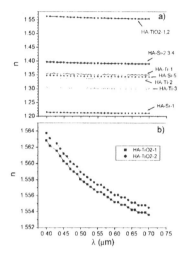

Figure 8 – Spectral dependence or refractive index n of a few sol-gel derived HAp films with different porosities (void concentration).

The use of a metallic Ti buffer does not change essentially the film porosity, while depositions on TiO₂ buffer layer can result in highly compacted HAp layers. For thinner (39nm) and more porous TiO₂ buffer layer, the void concentration is reduced to 5%, while for thicker buffer layer, the porosity reaches values smaller than 1% (see Table I).

We can conclude that the porosity of sol-gel derived HAp films can be varied in a broad range from 60% to less than 1%, by changing the annealing temperature rate and/or the buffer nature.

CONCLUSIONS

Optimization studies have been performed for single crystalline phase HAp film formation. Specific annealing temperatures of the HAp formation at about 150°C, 250°C, 320°C and 510°C were detected using TG-DTA measurements on powders obtained by evaporation of precursor solution. This thermo-analysis, together with XRD and SEM investigations on massive HAp powder samples has offered important information for optimization of HAp films. Additional to XRD and SEM studies on HAp films, the temperature evolution of the FTIR spectra were used to investigate the annealing effects. Single crystalline phase of HAp films with different porosity was obtained by changing the annealing temperature rate and the nature of the substrate. SE was used not only to evaluate the optical constants of the films, but also the porosity. A TiO2 buffer layer is favorable for compact HAp film formation. Layers with void concentration of less than 1% as well as very porous films with 60% voids were obtained.

ACKNOWLEDGEMENTS
The authors would like to thank G. Aldica for thermo-gravimetric measurements. The financial support of CEEX programs, contract No. 318 /2006 is gratefully acknowledged.

REFERENCES
[1]D.B.Haddow,P.F.James, R. van Noort, Sol-gel derived calcium phosphate coatings for biomedical applications, J.Sol-gel Sci.Technol., 13, 261-265 (1998).
[2]K. De Groot, Bioceramics consisting of calcium phosphate salt, Biomaterials 1, 47-50(1980).
[3]P.Li et al., J.Am. Ceram. Soc.,75, 2094 (1992).
[4]J.Wen, Y.Leng, J.Chen, C.Zhang, Chemical gradient in plasma sprayed HA coatings, Biomaterials 21, 342-350 (2001).
[5]J-S Chen, H-Y Juang, M-H Han, Calcium phosphate coating on titanium substrate by a modified electrocrystallization process, J. Mater Sci:Mater Med 9, 297-300 (1998).
[6]F.J. Garcia-Sanz, M.B. Mayor, J.L. Arias, J. Pou, B. Leon, M. Perez-Amar, Hydroxyapatite coatings: a comparative study between plasma –spray and pulsed laser deposition techniques, J.Mater.Sci:MaterMed 8, 861-865 (1997).
[7]J.G.C. Wolke, K, Van Dijk, H.G. Schaen, K. De Groot, A. Jansen, Study of the surface characteristics of magnetron- sputter calcium phosphate coatings, J. Biomed. Mater Res. 28, 1477-1484 (1994).
[8]A. Montenero, G. Gnappi, F. Ferari, M. Cesari, M. Fini, Sol-gel derived hydroxyapatite coatings on titanium substrate, J. Mater. Sci. 35, 2791-2797 (2000).
[9]W. Weng, G. Shen,G. Han, Low temperature preparation oh hydroxyapatite coatings on titanium alloy by a sol-gel route, J. Mater. Sci. Lett., 19, 2187-2188 (2000).
[10]E. Tkalcec , M. Sauer, R. Nonninger, H. Schmidt, Sol-gel derived hydroxyapatite powders and coatings, J.Mater. Sci., 36, 5253-5263 (2001).
[11]C. You , S. Oh, S. Kim, Influences of heating condition and substrate-surface roughness on the characteristics of sol-gel-derived hydroxyapatite coatings, J. Sol-gel Sci. Technol., 21, 49-54 (2001).
[12]A. Jillanevska, D. T. Hoelzer, R. A. Condrate, An electron microscopy study of the formation of hydroxyapatite through sol-gel processing, J. Mater. Sci., 34, 4821-4830 (1999).
[13]M. Yoshimura, H. Suda, K. Okamoto, K, Ioku, Hydrothermal synthesis of biocompatible whiskers, J. Mater Sci, 29, 3399 (1994).
[14]D-M Liu, T. Troczynski, W. J. Tseng, Water-based sol-gel synthesis of hydroxyapatite: process development, Biomaterials 22, 1721 (2001).
[15]I-S. Kim, P.N. Kumta, Sol-gel synthesis and characterization of nanostructured hydroxyapatite powder, Mater. Sci.Eng., B 111, 232 (2004).
[16]T. A. Kuriakose, S. N. Kalkura, M. Palanichamy, D. Arivuoli, K. Dierks, G. Bocelli, C. Betzel, Synthesis of stoichiometric nano crystalline hydroxyapatite by ethanol-basedsol–gel technique at low temperature, J. Cryst. Growth, 263, 517–523 (2004).
[17]H-W Kim, J. C. Knowles, H-E Kim, Improvement of hydroxyapatite sol-gel on titanium with ammonium hydroxide addition, J. Am. Ceram. Soc., 88, 154-159 (2005).
[18]H-W Kim, Y-H Koh, L-H Li, S. Lee, H-E Kim, Hydroxyapatite coating on titanium substrate with titania buffer layer processed by sol–gel method, Biomaterials, 25, 2533–2538 (2004).
[19]X. Nie, A. Leyland, A. Matthews, Deposition of layered bioceramic hydroxyapatite/TiO2 coatings on titanium alloys using a hybrid technique of micro-arc oxidation and electrophoresis, Surface and Coatings Technology, 125, 407–414 (2000).
[20]A. Balamurugan, G. Balossier, S. Kannan, J. Michel, S. Rajeswari, In vitro biological, chemical and electrochemical evaluation of titania reinforced hydroxyapatite sol–gel coatings on surgical grade 316L SS, Mat. Sc. Eng. C 27, 162–171 (2007).
[21]D. A. G. Bruggeman, Berechnung verschiedener physikalischer Konstanten von heterogenen Substanzen, Ann. Phys., 24, 636 (1935).

EVALUATION OF AGGREGATE BREAKDOWN IN NANOSIZED TITANIUM DIOXIDE VIA MERCURY POROSIMETRY

Navin Venugopal, Richard A. Haber

ABSTRACT:

Six nanosized titanias were investigated for aggregate breakdown. The powders all commonly exhibited particle (aggregate) sizes of approximately 1 micron via light scattering and primary particle sizes of approximately 20 nm via BET ESD estimation yet the powders appeared to show strongly different rheological characteristics. Investigation of the aggregate strength via compaction curves indicated yield points on the order of 500 MPa as compared with 4-20 Pa for suspension yield strength. Three distinct aggregate size regimes of size orders of approximately 100 microns, 10-20 microns and 1 micron were found with the variation in strengths observed corresponding to an attack on a different aggregate size regime. Investigations via mercury porosimetry found peaks for tape cast samples at ~250-400 nm corresponding to intact 1 micron aggregate units. Peaks in the nanosized regime were found corresponding to loose coordination of 20 nm primary particles. Investigation of pressed pellets from compaction curves found the absence of this peak and the shift of nanosized peaks to values from 5-10 nm suggesting the attack of the 1 micron aggregated unit and the rearrangement of the 20 nm units.

INTRODUCTION/BACKGROUND:

The use of nanomaterials to form bulk shapes has gained significant attention due to their offering unique properties with regards to rules established for conventional materials as well as their suprerior inherent advantages for variables such as diffusion length, specific surface area and particle number density. A common drawback of materials exhibiting this length scale is the large amount of fluid vehicle required to facilitate their processability by conventional techniques including paste or slurry formation. This is further compounded by particle number density proliferation at this length scale, smaller separation between particle surfaces and exacerbated particle collision frequency. Consequences of this include undesired aggregation into micron-sized 'effective units' denying the advantages afforded at the nanoscale[1,2].

Aggregates and similar terms such as agglomerate, floc or coagulate are terms that characterize a random assemblage of constituent particulate subunits[3]. The models available to consider the aggregation behavior materials assess strength of interaction in specific scenarios, such as the network model of particulate suspensions which associates rupture of aggregate 'bridges' with the onset of fluid flow[4,5]. An additional technique to characterize the strength of aggregates is suggested by Niesz in producing powder compaction diagrams whereupon the extrapolation of linear regions of Stage I and Stage II compaction's intersection provides a measure of aggregate yield strength[6].

The loose and disordered arrangement of aggregates prevents a characterization via conventional packing models while still intact. In the aforementioned two techniques, aggregate rupture should inevitably result in a rearrangement of the constituent subunits from compaction pressure above the yield point or shear stresses larger than the hydrodynamic stresses imparted to rupture the network in suspension. Upon rearrangement, the system will reconfigure according to a specific model such as those described initially by White and Walton for spherical particles[7].

Further to this is the apparent correlation between particle size and the size of the interstices produced under certain packing structures[8].

In instances of a simple aggregate system with one iteration of aggregation, this is trivial as there are only two configurations that can exist corresponding to intact aggregates of randomly coordinated primary particles and free flowing primary particles. The situation becomes nontrivial when multiple aggregation iterations are encountered. This has been reported in systems such as the work of David et al.[9] synthesizing nanosized ZnS where up to four separate and independent iterations of aggregation are reported. Their four iterations were reported to be of specific sizes and were strongly a function of specific synthesis conditions.

In the processing of nanosized materials which exhibit multiple aggregation iterations, it may be argued that certain aggregation iterations may be targeted for preservation depending on the application targeted for the system. This investigation intends to detail efforts at measuring the stresses required to rupture a series of nanosized titanium dioxide powders synthesized from a sulfate process. Specific stresses will characteristic of the stresses required to rupture certain iterations of aggregation. Tape casting will be utilized as a means of evaluating a controlled shear regime and its effect on aggregate breakdown in a formed part. This will be furthered by the use of mercury porosimetry to measure the interstices produced in a specific process as a means of back-calculating the size of the unit flowing under the specified stress regime.

EXPERIMENTAL

The titania utilized in this investigation was synthesized from a sulfate process, illustrated in Figure I. Modification of synthesis variables such as seeding level, washing volume or calcinations temperature were found to affect starting powder characteristics, specifically the soluble sulfate level and specific surface area.

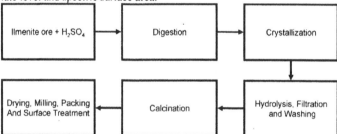

Figure I: Flowchart of the sulfate process for production of titanium dioxide

Tape Casting

Tape casting was performed via an air-driven motor with a moving doctor blade. The doctor blade was manufactured by and obtained from Richard Mistler Inc.. Doctor blade height was adjusted via micrometers attached to the doctor blade. Casting was carried out on a glass substrate, onto which a Mylar film was placed. Doctor blade motion was attained by placement on a chain connected to the air-motor assembly. Variation of the air pressure resulted in variation in the chain velocity and ultimately the casting speed. Pressures of 20 and 55 psi were found to correspond to casting velocities of 0.85 cm/sec and 9.09 cm/sec. A small quantity of slurry was placed ahead of the doctor blade. The slurry was composed of Deionized water and titania with no additional additives or surfactants employed. Upon casting the tape was dried overnight and then calcined at 600 degrees Celsius overnight.

Powder Compaction

For compaction curves, the powder was weighed as a 0.140 g sample ± 0.005 g and compacted using a stainless steel cylindrical KBr pellet die with a cross-sectional diameter of 12.7 mm (0.5 inches). All compaction was performed via an Instron 4505 loading frame using a 100 kN load cell. The load cell was calibrated and balanced prior to testing. The crosshead was lowered until a near-zero gap was achieved between the crosshead and the top surface of the punch. Compaction loading was performed at a velocity of 1.8 mm/min until the maximum load of 100 kN was achieved, whereupon the computer-controlled crosshead automatically was stopped. Compaction unloading was carried out at 1.8 mm/min immediately following cessation of the load procedure and proceeded until the crosshead was visually observed to no longer be in contact with the top punch. Upon ejection of the sample from the die, the sample thickness was measured using calipers and the mass was measured. The stage I/stage II transition was calculated by linear regression[10].

Mercury Porosimetry

Packing characteristics of the resultant tape were assessed via mercury porosimetry. Each sample was dried for 24 hours prior to testing. For all mercury porosimetry, samples were dried for 24 hours at 110 degrees Celsius. Samples were then placed in 3 cc bulbs that were evacuated to 50 □m Hg pressure for 5 minutes before being filled with mercury. High pressure analysis was performed via a Micromertics 366 Porosimeter for applied pressures ranging from 0.5 to 30,000 psi.

RESULTS:

Particle size distribution is presented in Figure II, and surface area is presented in the Table I. Powder density was previously measured via helium pycnometry and found to be 3.84 g/cm^3 for all powders. The light-scattering particle size distribution data show little variation in the starting powders size distribution or median particle size. Median particle diameters are centered at approximately 1 □m and appear to exhibit a log-normal distribution extending into the submicron range. The light-scattering data, however, offer few insights into the system without further characterization via other techniques. Such information can be obtained via the multi-point BET surface area measurements provided. The surface areas for the powders investigated in this study ranged from 71 to 127 m^2/g. Using Equation I below in combination with density data provided and shown in Table I, the equivalent spherical diameter (ESD) was calculated and found to range from 12 to 22 nm.

$$ESD_{BET} = \frac{6}{SSA \times \rho_{particle}} \quad (1)$$

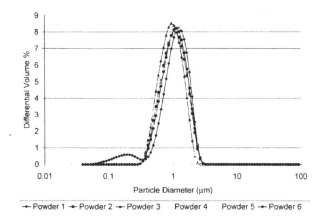

Figure II: Particle size distributions of the powders investigated via light scattering

Table I: A summary of powder characteristics and computed quantities

Powder	d_{50} (μm)	BET (m²/g)	Calculated ESD (nm)	Powder Density (g/cm³)
1	1.26	89.37	17	3.84
2	1.05	71.64	22	3.84
3	0.95	71.53	22	3.84
4	1.26	126.88	12	3.84
5	1.15	110.22	14	3.84
6	1.15	117.18	13	3.84

The dry powders were investigated via scanning electron microscopy and were found to exhibit three distinct aggregation phases as seen in Figure III a)-c). These three iterations of aggregation were found to exist as three separate size regimes. As indicated in Figure III(a), the system appears to initially exhibit primary particle sizes on the order of tens of microns, confirming the approximate order of magnitude given via estimation of ESD via BET. The clustering of primary particles results in a primary scale aggregate approximately 1 μm in diameter, corresponding to the peaks exhibit in light-scattering particle size analysis.

In Figure III(b) it can be seen that the primary scale aggregates themselves appear to cluster into a secondary scale aggregate approximately 5-10 μm in diameter; in Figure III(c), it appears that the secondary scale aggregates form a tertiary scale aggregate of diameter 100 μm and greater. It appears that the aggregates and the primary particles are of ill-defined shape, and cannot be conveniently described by a specific particle shape. This prevents the direct fit to a specific particle packing model of a particular shape. Due to the ill-defined shape and the apparent lack of a specific aspect ratio so as to cause a deviation from a shape factor of 1.0, approximations in consideration of particle packing for each aggregation iteration will utilize spherical models

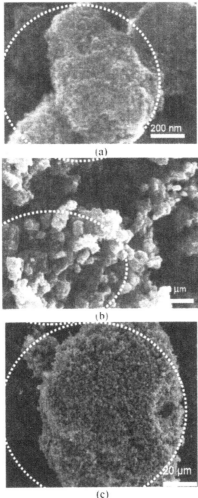

(a)

(b)

(c)

Figure III: Multiple Aggregation Stages seen via Scanning Electron Microscopy for Powder 1

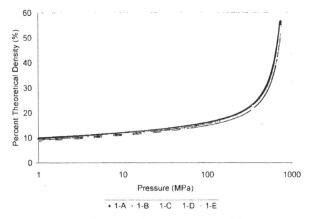

Figure IV: Sample compaction curves generated for Powder 1

Powder compaction was performed on each of the powder samples to establish upper limit boundary conditions for stability of the aggregate. A sample compaction curve for each of the six powders is shown in Figure IV with linear regression results plotted in Table II. In all instances of compaction it can be seen that the pressure utilized is sufficient to produce what appears to be a transition between Stage I and Stage II of the typical compaction curve. There appears to be no transition to Stage III evident suggesting that 750 MPa is insufficient pressure to produce rearrangement of the primary particles believed to occur in Stage III. It is possible that the system's primary particles possess a sufficiently high strength to withstand deformation at this pressure. It is also possible that the granule rearrangement has not been fully optimized to proceed to the next stage of compaction. However, the pressure utilized appears to be sufficient to see a transition to Stage II and subsequently to fragment the aggregates.

The compaction results appear to indicate aggregate breakdown as evidenced by the transition from Stage I to Stage II compaction at pressures of approximately 500 MPa suggesting this as the strength of primary scale aggregates.

Table II: Calculated Yield Points in Compaction

Powder	Mean Yield Point (MPa)	Standard deviation
1	512	6.64
2	501	2.67
3	501	6.16
4	499	2.09
5	503	7.67
6	485	7.03

Figures V and VI show results of mercury porosimetry performed on tape cast pieces of low and high velocity respectively. These plots indicate that there does not appear to be a significant amount of aggregate breakdown attained with increasing the casting velocity by 1 order of magnitude. The porosimetry on the tapes indicates that there are three peaks produced, corresponding to three diameters commonly exhibited in the tapes of the powders. The first peak, for a pore diameter of 212 μm. corresponds to residual pores not fully eliminated via shear; the peak is believed to be misleadingly significant since the size of the pore diameter results in a significant volume of mercury intrusion in spite of a relatively low number of pores exhibiting this size.

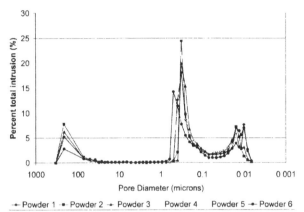

Figure V: Mercury porosimetry of tapes cast at 0.85 cm/sec

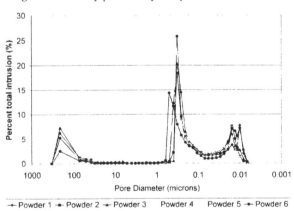

Figure VI: Mercury porosimetry of tapes cast at 9.09 cm/sec

The second major peak (and commonly the largest peak for tapes of low and high velocity for each of the six powders investigated) corresponds to a submicron pore diameter. The occurrence of this peak value ranges from pore diameters of 0.32 to 0.49 μm depending on the powder investigated. Varying casting velocities do not appear to affect the location of this peak for each of the individual powders. In particle packing, a relationship can be derived between the size of the interstices and the size of the particles assuming a roughly monomodal distribution. For spherical particles, typically this ratio varies between 0.22 for a tetrahedral configuration and 0.51 for cubic arrangements. Since the particle size is typically known, this technique is used to correlate the ratio measured to determine the particular packing model that a system corresponds to.

In this instance, however, the model is being used to confirm that the flowing unit in this process corroborates to a specific aggregation stage. It has been established that the primary scale aggregates are approximately 1 µm in size; if that is the flowing unit, the interstices resultant in a green body will range from 0.22 to 0.51 µm in diameter. The peak diameters measured from mercury porosimetry suggest that the major flowing unit in tape casting is the preserved primary scale aggregate since the major peaks fall within the aforementioned range of mercury intrusion vs. pore diameter. Moreover, it is notable that for Powders 4, 5 and 6, this peak occurs at 491, 390 and 390 nm respectively. Given the similarity in the d_{50} values observed for these powders in Table II, in spite of the strong variations in powder surface area it can be argued that these three powders will exhibit rougher surface characteristics and subsequently will exhibit a lower packing efficiency. Subsequently, since packing efficiency is related to the size ratio between the interstices and particles, it can be argued that the greater degree of aggregation of these three powders produces a more fractal and irregular surface to the particle. This may explain why the powders result in a lower packing efficiency resulting in larger interstices resultant from rearrangement of the 1 µm unit for these powders.

A third peak is observed for pore diameters within the nanosized regime. This regime in some instances features multiple broad peaks suggesting a more disorganized assemblage at this length scale. It is speculated that these peaks reflect the loose assemblage of primary particles within a primary scale aggregate (referred to alternately as intraparticle porosity). The presence of these loosely defined peaks at this length scale suggests that tape casting does not attack the primary scale aggregates and the intraparticle porosity inherent in the powders upon synthesis is preserved.

By comparison, the pressed pellets have typically been inferred to attack primary scale aggregation given their use typically to measure granule strength in powder compaction. Figure VII plots the programs of the pressed pieces. Initially, the pellets also appear to exhibit a large peak at approximately 210 µm which is again believed to be an artifact of the low cohesive strength in the sample produced by the lack of a binder. However, the predominant difference seen in the pellet intrusion is the absence of the major submicron peak seen in the tapes. A minor peak is seen to occur at approximately 6 µm for the pellets which may correspond to interstices between tertiary and higher order aggregation stages. Incremental intrusion, however, results in one other significant peak, which occurs in the nanosized pore diameter range.

In the nanosized range there appear to be two distinct peaks which may correspond to two distinct states of aggregation. One peak is seen to occur between 10 and 20 nm, suggesting that some remnants of the intraparticle porosity are retained. However, a larger peak is seen commonly below 10 nm in all pellets. Compaction curves appeared to indicate a yield point at approximately 500 MPa during compaction to 750 MPa suggesting that the pellets were pressed to Stage II of compaction, where the primary particles begin to fill the interstices between the ruptured granules. The presence of a larger peak below 10 nm suggests that the individual 20-40 nm nanocrystallites are filling the voids caused by packing of the 1 µm primary scale aggregates.

Moreover, it is suggested that the presence of a peak below 10 nm corresponds to a denser packing of the individual primary particles, confirming the inferences drawn previously in compaction curves.

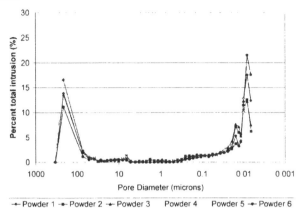

Figure VII: Mercury porosimetry of pellets pressed to 750 MPa

It can be argued that, for each of Figures V-VII that assumptions regarding monosized units may explain the results observed. It can be seen from as early as Figure II that it can be questioned how effectively it can be assumed that the monosized approximation holds for these systems. This may explain the overall width seen in the pore distributions believed to correspond to primary scale aggregates. Moreover, it can be argued that even if the spherical monosized assumption is reasonable with this system it is likely that the non-spherical nature of the aggregates and the primary particles make the random close packed model to be the more appropriate model for consideration. However, because of the nature of the random close packed model no specific interstice to particle ratio can be acquired.

To resolve this, it is argued that since the random close packed model yields a packing efficiency that is intermediary with respect to the aforementioned cubic and tetrahedral models (64%), it can be inferred that while there is an expected distribution of pore sizes resulting from coordination of units of a specific size, these pore sizes will also be intermediary with respect to the ratios of the cubic and tetrahedral models. In consideration, the pore sizes observed via mercury porosimetry for those corresponding to sizes attributed to primary scale aggregate reordering fall within the expected values[8].

It should be noted for the purpose of this investigation that the particle packing models rely on assumptions of an approximately spherical configuration or packing configurations based on particles exhibiting similar aspect ratios to spheres. Based on qualitative observations via SEM the particles do not appear to exhibit specific anisotropy for a particular length scale (if exhibiting any specific shape at all) so it is believed to be reasonable to use the spherical models. Particles of high anisotropy may pack in configurations whereupon the ratio of the particle size to the resultant interstices upon alignment may be significantly smaller. Reed specifically contends that angular particles or particles exhibiting anisotropy of this nature will randomly occupy 50-60% of the volume[8].

SUMMARY:
Sulfate-processed titania powders of high specific surface area appear to exhibit multiple aggregation iterations. The residual sulfate in the system, established to be soluble sulfate via spectrophotometry, appears to serve as an indifferent electrolyte in the viscoelastic suspension. The presence of a greater amount of indifferent electrolyte results in a smaller electrical double layer thickness subsequently reducing the degree of double layer overlap with other powder particles and reducing the degree of networking in the suspension. Compaction curves appear to indicate that the strength of primary scale aggregates range from 484 MPa to 511 MPa. Bulk forming via tape casting appeared to preserve primary scale aggregates as seen qualitatively via scanning electron micrographs. This appears to be corroborated by mercury porosimetry indicating the presence of submicron peaks ranging from 200-500 nm which suggest interstices formed from approximately 1 μm units via particle packing models for spherical particles. Mercury porosimetry of the compacted pellets appeared to corroborate the rupture of primary scale aggregates based on the absence of the aforementioned submicron peak. Additionally, the nanosized peaks additionally exhibited in tape casting appear to have shifted to smaller sizes for compacted samples which indicate rearrangement of primary particles upon rupture of the primary scale aggregate.

REFERENCES
1. Nanomaterials: Synthesis, Properties and Applications, Edited by A.S. Edelstein and R.C. Cammarata, (Institute of Physics Publishers, Philadelphia PA, 1998)
2. N. Kallay, S. Zalac, "Stability of Nanodispersions: A model for Kinetics of Aggregation of Nanoparticles", *J. of Coll. and Interface Sci.* **253** 70-76 (2002)
3. V. A. Hackley, "Guide to the Nomenclature of Particle Dispersion Technology", *NIST Special Publication 945*
4. T.F. Thadros, *Rheology of Unstable Systems*, Industrial Rheology Lecture Notes (The Center for Professional Advancement, 1994)
5. H. Wu, M. Morbidelli, "A Model Relating Structure of Colloidal Gels to Their Elastic Properties," *Langmuir*, **17**, pp. 1030-1036 (2001)
6. D.E. Niesz, R.B. Bennett and M.J. Snyder, "Strength Characterization of Powder Aggregates," *Am. Ceram. Soc. Bull.*, **51** [9], 1972 p.677
7. H.E. White, S.F. Walton, "Particle Packing and Particle Shape," *J. Am Ceram. Soc.*, **20** pp. 155-166 (1937)
8. Reed, J.S., Principles of Ceramic Processing, 2nd ed. (John Wiley & Sons Inc., New York, NY, 1995) p. 218-227
9. R. David, F. Espitalier, A. Cameirao, L. Rouleau, "Developments in the understanding and modelling of agglomeration of suspended crystals in crystallization from solutions," *KONA Powder and Particle*, **21**, 40-53 (2003)
10. P.R. Mort, R.Sabia, D.E. Niesz, R.E. Riman "Automated generation and analysis of powder compaction diagrams" *Powder Technology* **79** 111-119 (1994)

ENRICHMENT AND VACUUM-SINTERING ACTIVITY OF COLLOIDAL CARBON SUBMICRO-SPHERES

Jianjun Hu, Zhong Lu, Qiang Wang*
Department of Materials Science and Engineering, Jiangsu Polytechnic University and
The Key Laboratory of Polymer Materials of ChangZhou City
Changzhou,Jiangsu, 213016, PR China

ABSTRACT

Colloidal carbon submicro-spheres with the diameter of 100-200nm can be formed by a glucose-thermal method at 180 °C for 4 hours. Different electrolytes were tested to separate carbon spheres from the reacted system. The vacuum-sintering activity at 700°C,1500°C was explained by the intrinsic muti-functional groups surface of these carbon spheres. The measured BET value of carbon spheres before and after vacuum sintering was 26.6m^2/g, 444.3m^2/g and 65.8m^2/g, respectively.

INTRODUCTION

Since the discovery of fullerene, carbon nanotube and success commercial application of mesoporous carbon micro-spheres (MCMB) in Li-ion battery, nano-structured carbon materials have attracted tremendous attention from both academic and industrial fields for their potential applications, such as adsorbents, catalyst supports, nano-composite, energy storage media and etc[1]. Many efforts were exerted to fabricate globular form of carbon particles which includes arc-discharge[2], CVD[3], micro-emulsion, and reduction of different carbon precursors by metals[4-5]. Solution-based route to synthesis carbon nano-particles from biomass precursor was proved to be an effective method for its ability to control size and geometry precisely[6]. However, separation of solid products from the colloidal suspensions really is the key problem. High speed centrifuge is routinely applied, but it is not a low energy-cost choice. Herein, we report an electrolytes-assisted separation of colloidal carbon spheres from the carbonized glucose system. Also, the vacuum-sintering behavior and the structure evolution of these carbon spheres are investigated.

EXPERIMENTAL SECTION

In a typical procedure, 10g glucose (AR grade, Sinopharm chemical Reagent Co, Ltd.) was dissolved in 100ml distilled water, and then appropriate volume of solution was transferred to a stainless steel reactor with Teflon inner. It was heated up to 180°C at a rate of 8°C/min and kept for 4 hours. The pressure inside of reactor is about 10 atm, according to the phase diagram of pure water. After it was cooled down to ambient temperature, a dark brown suspension turned out. Electrolytes with different cation valancies (+1,+2 and +3) were dropped to the suspension ,respectively. Particle deposition could be observed apparently at different time. By removing the upper solution and drying the solid product at 110°C in a glass flask, colloidal carbon spheres were obtained. Some of these spheres was put in a crucible made of graphite and heated at 700°C and 1500°C under the condition of vacuum (the average degree of vacuum could be as high as 5*10^{-3}Pa). The carbon spheres before and after sintering were characterized by powder XRD (Cu Kα radiation D/max2500PC,Rigaku), SEM (JSM6360, JEOL), FT-IR(Nicolet-460, Hewlett Packard), N$_2$ sorption isotherms and BET were measured using a Micromeritics ASAP2010MC analyzer at 77 K. Before the measurements, all samples were degassed overnight at 523 K in a vacuum line.

RESULTS AND DISCUSSIONS

Colloidal or nano-particles synthesized in solution-based route have large surface area and the capability to absorb external charges which always leads to the stable colloidal suspensions. For example, the carbon suspension produced in our experiment can be existed for several months without any segment can be observed. Although centrifuge operation is used generally in solid-liquid phase separation, it still has to face it limitation in extreme diluted system or ultra-fine nanoparticles. In our experimental procedure, even the rotating rate of centrifuge machine is as high as 5000r/min, most of solid products can not be separated. We have to look for another strategy of destroying the stability of suspensions by adding appropriate electrolytes. Figure.1 depicts the curve of the concentration of different electrolytes and the total time of complete sedimentation. It can be seen that the total time of complete sedimentation decreases with the increasing concentration of electrolyte and when the concentration of electrolytes reach certain values, the consumed total time decrease significantly, for example, 3h for 10 g/l $MgCl_2 \cdot 6H_2O$,2h for 10g/l $AlCl_3 \cdot 6H_2O$ and 10h for 40g/ml NH_4HCO_3. This result is consistent with the prediction of DLVO theory, which suggests the interplay of Van der Walls forces and electrostatic double layer interaction induces the charge-stablized colloidal suspension, and the sensitivity of aggregation and deposition rates to ion strength or pH values of solution[7]. From the rate of sedimentation, it is reasonable to elucidate that the colloidal carbon submicro-spheres produced in our process is negatively charge. However, our result does not follow Schulze-hardy rule strictly, which might be caused by the ion effects of the complex intermediate compounds. It is worthy to note that all solid carbon spheres can be collected by this method.

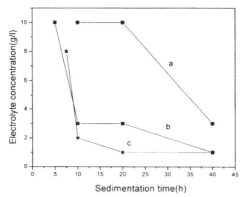

Figure 1. The curve of the concentration of different electrolytes and the total time of complete sedimentation.
a: NH_4HCO_3 as electrolyte.
b: $MgCl_2 \cdot 6H_2O$ as electrolyte.
c: $AlCl_3 \cdot 6H_2O$ as electrolyte.

Fig. 2 shows the SEM image of the submicro-spheres using NH_4HCO_3 as electrolyte without centrifugation operation (unless otherwise stated, all the characterization employed this sample). It was clearly demonstrated that the majority of the products exhibited perfect spherical morphology and the spheres varied from 100nm to 200nm in diameter in Figure a and b. No NH_4HCO_3 crystals can be identified in this figure for it was decomposed to gaseous NH_3, H_2O and CO_2 at drying temperature. Fig.2c and 2d are SEM image of the carbon spheres after they were carbonized in vacuum for 3h at 700°C and 1500°C,respectively. From Fig. 2c, it was found that the treated spheres tend to shrink in comparison with Fig.2b. It also can be seen that carbon micro-spheres structure as sintered microparticles formed by fused highly agglomerated carbon spheres in Figure 2d.

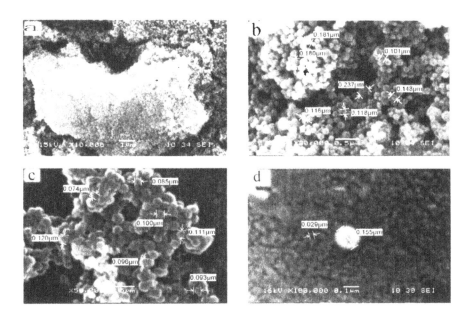

Fig .2   SEM photographs of carbon submicro-spheres.
a: SEM image of as-prepared carbon submicro-spheres at low magnification .
b: SEM image of as-prepared carbon submicro-spheres at high magnification.
c: SEM image of carbon submicro-spheres sintered at 700°C.
d: SEM image of carbon submicro-spheres sintered at 1500°C.

Shown in Figure 3 are the XRD patterns for the as prepared sample(a) and the same sample subjected to additional graphitization at 700°C and 1500°C(b,c). There is only a broad diffraction in Figure3a of as prepared carbon spheres, which means it is completely amorphous to X-ray. In Figure3b and 3c, there exists two broad diffraction peaks at about $2\theta=23°$ and $44°$, which could be indexed as (002) and (100) diffraction peaks of turbostratic carbon[8]. From the relative intensity of these two peaks, it can be expected that high temperature is favorable for the degree of graphitized of these carbon spheres. The distance of inter-spacing (d_{002})between layers of sample b and sample c is 0.3864 and 0.3705 nm, calculated according to a JSPS procedure[9], which also indicates the distorted (002) planes in contrast with high crystalline graphite.

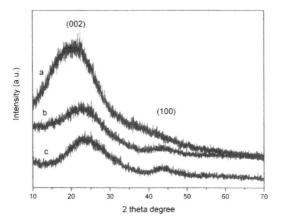

Figure. 3 XRD patterns of as-prepared and treated carbon submicro-spheres.
a: XRD pattern of as-prepared carbon submicro-spheres.
b: XRD pattern of carbon submicro-spheres treated at 700°C.
c: XRD pattern of carbon submicro-spheres treated at 1500°C.

The FT-IR spectrum (Fig.4) was used to investigate the residual groups present on the surface of carbon particles, and the following absorption bands can be observed: band at 3400cm^{-1} due to the stretching vibration of O-H bond, bands at 2926cm^{-1} due to the stretching vibration of C-H bond, band at 1620 and 1512cm^{-1} due to C=C and C-C bond of aromatic phenyl ring. The bands at 1702 and 1250cm^{-1},which are attributed to the stretching vibration of C=O bond, specifically of ketone, aldehyde and carboxylic acid functionalities, and C-OH stretch vibration[10], respectively.All of those bands reveal that there are large numbers of functional groups such as –OH and C=O groups on the original sample surface[11-12]. Fig. 4b illustrate the FT-IR spectrum of the carbon submicro-spheres heat-treated at 700°C. Most of above absorption band disappeared, only broad absorption peak at 3400cm^{-1} of adsorbed water and two weak peaks at 2925 and 2857cm^{-1} which were assigned to hydrogen bonded sp^3 carbon were kept[13]. The same result was obtained from Fig.4c at 1500°C sintering. Furthermore, the disappear of these functional groups is accordance with the above SEM observation. High temperature sintering should facilitate the surface chemical reaction between these functional groups and induce fusion and connection of carbon spheres.

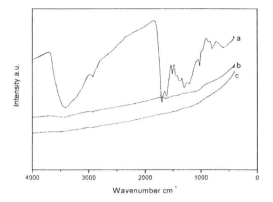

Figure.4   FT-IR spectra of carbon submicro-spheres
a: as-prepared carbon submicro-spheres.
b: 700°C sintered carbon submicro-spheres .
c: 1500°C sintered carbon submicro-spheres .

Table.1 shows the specific BET surface area of the carbon submicri-spheres measured by Nitrogen-adsorption method before and after vacuum sintering. Upon vacuum sintering at 700°C for 3 hours, BET surface area was increased by about 444.3m²/g from its original 26.6 m²/g. This increase was ascribed to the as-formed accessible nano-pores of average size around 10-20Å by BET calculation. Many authors suggested that these pores can result from the reaction of hydrogen and oxygen which yields water, and which helps in formation of very small pores[14]. However, with the temperature increasing to 1500°C, the BET surface decreased to 65.8m²/g again. This result should be explained by the above micro-pores had been collapsed in high temperature sintering, Tanaike *et al* reported similar result in mesoporous carbon from defluorinated PTFE[15].

Table.1 The specific surface area of carbon submicro-spheres treated at different temperature.

Sample	S_{BET} (m²/g)
As-prepared carbon submicro-spheres	26.6
Carbon submicro-spheres of 700°C treated	444.3
Carbon submicro-spheres of 1500°C treated	65.8

CONCLUSIONS

Electrolytes with different cation valances were employed to separate solid carbon spheres in glucose- precursor reaction system. The choice of easy to decompose electrolyte such as NH_4HCO_3 is favor of keep the purity of products. Cheap raw materials and a low cost separation process make this technique of industrial importance.

ACKNOWLEDGMENTS

We acknowledge the financial supports from Jiangsu Polytechnic University and Changzhou city (grant of K2007304).

REFERENCES

[1]M.S. Dresselhaus, G. Dresselhaus, P.C. Eklund, Science of Fullerenes and Carbon Nanotubes, Academic, San Diego,1-14 (1996).

[2]S. Iijima, T. Ichihashi, Single-shell carbon nanotubes of 1-nm diameter, Nature,**363**, 603-05 (1993).

[3]J.Y. Miao, D.W. Hwang, C. C. Chang, Uniform carbon spheres of high purity prepared on kaolin by CCVD, Diamond and Related Materials, **12**, 1368-72 (2003).

[4]Z. S. Lou, C.L. Chen, Q. W. Chen, and J. Gao, Formation of variously shaped carbon nanotubes in carbon dioxide–alkali metal (Li, Na) system, Carbon, **43**,1103-08 (2005).

[5]Q. Wang, F. Y. Cao, Q. W. Chen, Formation of carbon submicro-sphere chains by defluorination of PTFE in a magnesium and supercritical carbon dioxide system, Green. Chem,**7**.733-36 (2005).

[6]X. M. Sun, Y. D. Li, Colloidal Carbon Spheres and Their Core/Shell Structures with Noble-Metal Nanoparticles, Angew. Chem. Int. Ed.**43**,597 –601 (2004).

[7]Z. Adamczyk, P. Weronski, Application of the DLVO theory for particle deposition problems, Advances in Colloid and Interface Science, **83**, 137-226 (1999).

[8]S. B. Yoon, G. S. Chai, S. K. Kang, Graphitized Pitch-Based Carbons with Ordered Nanopores Synthesized by Using Colloidal Crystals as Templates, J. Am. Chem. Soc., **127**, 4188-89 (2005).

[9]Japan society for the promotion of science (117th committee). On the measurement of lattice parameters and unit cell dimension of artificial graphite,A.Tanso **36**,25-34 (1963).

[10]Drbohlav J, Stevenson WTK, The oxidative stabilization and carbonization of a synthetic mesophase pitch, part II: The carbonization process, Carbon, **33**,713–31 (1995).

[11]Z.F. Wang, P. F. Xiao, N. Y. He, Synthesis and characteristics of carbon encapsulated magnetic nanoparticles produced by a hydrothermal reaction, Carbon.**44**.3277-84 (2006).

[12]Silverstein RM, Bassler GC, Morrill TC, In: Spectrometric identification of organic compounds, John Wiley, New York,107–24 (1991).

[13]F. Y. Cao , C. L. Chen , Q. Wang, and Q. W Chen. Synthesis of carbon–Fe_3O_4 coaxial nanofibres by pyrolysis of ferrocene in supercritical carbon dioxide, Carbon, **45**, 727–31 (2007).

[14]V. G. Pol, M. Motiei, A. Gedanke, and J. Calderon-Moreno, Carbon spherules: synthesis, properties and mechanistic elucidation, Carbon,**42**, 111–16 (2004)

[15]O.Tanaike, H. Hatori, Y.Yamada, and S. Shiraishi, Preparation and pore control of highly mesoporous carbon from defluorinated PTFE, Carbon, **41**, 1759-76 (2003)

NITROGEN DOPED DIAMOND LIKE CARBON THIN FILMS ON PTFE FOR ENHANCED HEMOCOMPATIBILITY

S. Srinivasan, Q. Yang[*]
Department of Mechanical Engineering, University of Saskatchewan, 57 Campus Drive, Saskatoon, SK S7N 5A9, Canada

V.N. Vasilets
The Research Center for Biomaterials, Research Institute of Transplantology and Artificial Organs, Shukinskaya 1, 123182 Moscow, Russia

Diamond like carbon (DLC) doped with nitrogen thin films were synthesized on silicon and Polytetrafluroethylene (PTFE) substrates using methane, argon and nitrogen mixture ion beams in a dual ion beam deposition (DIBD) system and hot wire plasma sputtering of graphite in argon and nitrogen gas mixture. The chemical composition, electronic structure, morphology, tribological, and hemocompatible properties of the DLC films were investigated by various methods including X-ray Photoelectron Spectroscopy, Scanning Electron Microscopy, Raman Spectroscopy, scratching, nanoindentation, ball-on-disc wear tests, and platelet adhesion technique. The effects of nitrogen doping and ion beam energy on the tribological properties of DLC thin films were studied. The results show that ion deposited films are dense and smooth and that nitrogen doped DLC thin films exhibits much lower friction coefficient and enhanced hemocompatibility. Nitrogen doping improved the hardness and hemocompatibility of DLC thin films.

INTRODUCTION

DLC thin films exhibit many unique properties like high hardness, high chemical inertness, high wear and corrosion resistance, excellent biocompatibility and very low coefficient of friction, which make this material ideal as protective, corrosion resistant and anti-wear coatings for a variety of applications including bearings, automotive components and biomedical implants[1-3]. DLC is an amorphous carbon with high content of sp^3 hybridization. Its amorphous nature make it feasible to doped with small amounts of different elements like N, O, F, Si and their combinations to improve and control their mechanical, tribological and biological properties of the DLC matrix. In the past decade, there is growing interest in studying nitrogen doped (N-doped) DLC coatings due to their remarkable field emission properties and possible applications as a cold cathode in vacuum microelectronic devices and in flat panel display technologies[4]. Low threshold electron emission from N-doped DLC thin films has been reported[4,5]. However, DLC films are always accompanied by a high internal compressive stress due to the ion bombardment during the deposition, which limits their possible applications. Recently, Silva et al.[6] investigated the effects of nitrogen doping on properties of DLC films and showed a clear reduction of the internal compressive stress without significant changes in mechanical hardness upon nitrogen doping.

Polymers possess many desirable properties for biomedical applications, including a density comparable to human tissues, good fracture toughness, resistance to corrosion, and ease of forming by molding or machining and are increasingly chosen as a material in a variety of implants for widespread applications. This trend will continue until tissue growth science and engineering reaches its maturity

---

[*] Corresponding author: qiaoqin.yang@usask.ca

level. For example, synthesized vascular grafts are now made mainly from extruded Polytetrafluroethylene (PTFE). The artificial small-vessels constructed from PTFE are the only alternative to autologous implants for more than 500,000 patients requiring coronary artery bypass surgery and more than 150,000 patients in need of lower limb surgical replacement operations. However, due to their limited hemocompatibility, blood clotting caused by blood-cell adhesion to PTFE becomes a serious problem when the diameter of vascular grafts made from PTFE is less than 6 mm. Thus, one of the most urgent problems to be resolved for wider applications of PTFE vascular grafts, and to avoid the need for patients to undergo repetitive surgery, is to improve hemocompatibility of PTFE-based prosthetic components.

DLC have been demonstrated to be biocompatible *in vitro* and *in vivo* in orthopedic applications. Coating polymer- based biomedical devices such as blood vessels, heart valves, and coronary stents with DLC thin films is expected to significantly improve their haemo- and bio-compatibility, tribological performance and lifetime. However, biomedical characteristics such as blood compatibility and thrombogenicity of DLC and their adhesion to respective polymers have not been fully investigated as required for practical implant applications.

In this paper, we report on the synthesis and characterization of N-doped DLC thin films using both ion beam deposition and plasma sputtering of graphite.

EXPERIMENTAL DETALS

1. Substrate materials

The substrate materials were Polytetrafluroethylene $-[CF_2-CF_2]_n-$ (PTFE) films (thickness 60 microns) and silicon (Si) wafers. The size of PTFE and Si samples was 20 mm×10mm. All samples were cleaned in ultrasonic bath with methanol and distillate water before deposition.

2. Deposition technique

2.1 Plasma sputtering

Plasma sputtering was conducted in a hot wire enhanced DC plasma reactor, as described in Ref. [7]. The PTFE samples were exposed to deposition in dc glow discharge plasma by sputtering a graphite target in the flow of Ar with 5% air. The gas flow and pressure in the plasma chamber were maintained at 25sccm and 5×10^{-2} Torr respectively. The square graphite target (25mm×25 mm) was located at 10 mm from round sample holder (diameter 50mm). The temperature of sample holder during deposition was measured by chromel / alumel thermocouple and ranged in 180 - 210 °C depending on plasma conditions. The temperature of hot wire was controlled by Micro Optical Pyrometer (Pyrometer Instr. Co., USA) and varied in the range 1950 – 2100 K depending on the filament current. The time of deposition ranged from 15 min to 3 hour. In the same experiment the polymer and Si wafer were deposited simultaneously.

2.2 Ion beam deposition

The ion beam deposition was performed in an IBAD system, schematically shown in Fig. 1. It consists of vacuum chamber made of stainless steel, a 12cm DC ion source suitable for etching and deposition, a sputtering target holder, a sputtering ion source, a shutter, a water cooled substrate holder that can be rotated and tilted to any degree. Carbon and metallic targets can be sputtered using the sputtering ion source. The substrate holder can hold specimens of 100 mm in diameter and can also be cooled with water to maintain the substrate close to room temperature during deposition and heated to required temperatures. The substrate can be rotated continuously to get a uniform deposition and tilted to an angle from 0° to 90° to control the ion bombardment incident angles. The typical experimental parameters are listed in Table 1.

3. Characterization

The surface chemical composition before and after deposition were analyzed by XPS spectroscopy (Kratos Axis Ultra spectrometer, USA). The XPS spectra were obtained with 90 degree take-off angle by using monochromatised Al K_α source. The charge neutralizer was on for all the analysis. The Raman spectra were obtained using a Reni Shaw micro-Raman System 2000 spectrometers operated at an argon laser wavelength of 514.5 nm. The laser spot size was around 2μm with a power of 20 mW. The surface morphology of the samples was examined by scanning electron microscopy (SEM) using secondary and backscattered electrons, 5kV accelerating voltage, and magnification up to 10000 (JSM T330, JEOL, Japan).The roughness of the thin films were measured using a profilometer.

Scratch and ball-on-disc wear tests were performed to examine the adhesive and tribological properties of the DLC films, respectively. The scratch tests were performed using a linearly increased load, from 2 to 15 N with a WC blade. A ball of 4 mm in diameter made of chrome steel was used in the friction and wear testing. The load used was kept constant from 0.5 N to 5 N. The morphologies of the wear track of DLC thin films were observed by optical microscopy. The experiments were conducted at room temperature and relative humidity of 30% to 50% without lubrication.

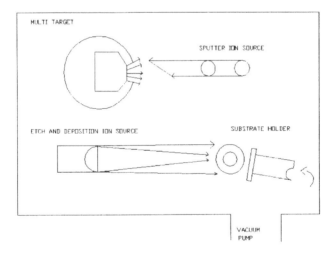

Fig.1 Schematic diagram of IBAD system

Table 1: Typical Experimental Conditions for Ion Beam Deposition

Parameters	Si wafer	PTFE
Base pressure (Torr)	4.20 E- 07	2.80 E- 06
Working pressure (Torr)	5.39 E- 04	7.08 E- 04
Ion beam energy (ev)	300	250
Ion beam current (mA)	75	75
Total gas (Ar+CH$_4$+N$_2$) (sccm)	34	28
Deposition time (hrs)	4	4
Deposition temperature	Room temperature	Room temperature
Coating thickness	150 nm	120 nm

4. Platelet adhesion technique.

The study of platelet adhesion was performed with informed consent of the donors. Five volunteers were included in the series of repeated experiments. Ten milliliters of blood was anticoagulated 1:9 with sodium citrate[8]. Platelet-rich plasma (PRP) was obtained by centrifugation of the whole blood at 100×g for 20 min at room temperature. The samples PRP drops (50 μl) were placed onto sample surfaces and incubated in humid atmosphere for different periods (from 5 to 30 min). The samples incubated with PRP for 15 min were chosen for further analysis. The number of platelets adhered to the surface during this time interval was appropriate for quantification, and the platelets did not form large thrombus-like structures. The samples were rinsed in normal saline to remove unadsorbed plasma proteins and weakly adhered platelets and then fixed in 2.5% glutaraldehyde and dehydrated in a series of ascending ethanols by the standard technique[8].

RESULTS AND DISCUSSION

1. Characterization of deposited film

Fig. 2 shows XPS spectra of PTFE film before and after deposition of carbon layer by the plasma sputtering. For the PTFE substrate, the XPS spectrum consists of F1s peak (688.2 eV) and less intensive C1s peak (291.4 eV). In the XPS spectrum of the film obtained after 15 min deposition one can see intensive C1s peak (285.0 eV) as well as O1s (532.1 eV) and N1s (399.8 eV) peaks. The important point is that no significant F1s peak was observed in the XPS spectrum of carbon film obtained after 15 min carbon deposition. It means that pin-hole-free carbon layer with the thickness more then 30 Å (depth of free electron path for 800 eV [9] is formed on the PTFE surface already after 15 min of graphite sputtering. The atomic surface concentrations calculated from measured integral intensities of XPS peaks for as deposited carbon film are: 72.2 at% C, 10.5 at% N and 14.0 at% O. The appearance of relatively high concentration of oxygen on the surface of carbon layer may be explained by the surface contamination and/or bonding of oxygen in post reactions with carbon free radicals formed during plasma deposition. After a few minutes of Ar$^+$ ion sputtering oxygen content decreased in three times while the nitrogen concentration only slightly changed. The chemical composition measured after ion etching was: 86.8at % C, 8.7at%N, and 4.5at%O. However, there is no oxygen in the ion beam deposited DLC films due to the high base vacuum and high purity of the gases used.

The SEM surface morphology of the DLC thin films deposited by ion beams, as shown in Fig. 3, reveals that the films are dense, smooth, uniform, and amorphous with no pores and discontinuity presented whereas carbon films synthesized by sputtering have a complex fibril-type 1-2-micron regular structure[7].

The surface roughness of the ion beam deposited DLC films, as listed in Table 2, was found to be ranging from Ra 1.15 nm to 1.4 nm whereas that of the plasma sputtering deposited carbon films ranges from Ra 11.4 nm to 12.6 nm. There was only a little bit increase in the surface roughness by the addition of nitrogen into DLC. The DLC films deposited by ion beams have a very smooth surface, much smoother than the carbon films deposited by the plasma sputtering [7].

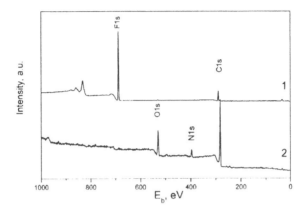

Fig. 2 XPS spectra of PTFE (1) before and (2) after the plasma sputtering deposition of carbon film

Fig. 3 Typical SEM micrograph of ion beam deposited DLC thin films

Table 2 Surface roughness of the ion beam deposited DLC thin films

Ion energy (ev)	DLC thin films deposited by ion beams			
	250		300	
Incident ion beam gas composition	75% Ar + 25% CH₄	70% Ar + 24 % CH₄+ 6 % N₂	75% Ar + 25% CH₄	70% Ar + 24% CH₄+ 6% N₂
Roughness (nm)	1.148	1.276	1.225	1.378

Raman spectra of the DLC thin films are shown in Fig. 4. The spectra taken from the ion beam deposited samples, either pure or N-doped, exhibits typical characteristics of DLC thin films[10], a broad peak at 1570 cm⁻¹ (G-band) with a shouldered peak at 1300 cm⁻¹ (D-band), whereas the spectra taken from the plasma sputtering deposited samples consists of two broad bands at 1575 cm⁻¹ and 1360 cm⁻¹, which are usually detected in amorphous carbon films[7-9]. According to Wagner's model[11] developed for Raman scattering of carbon, peak at 1575 cm⁻¹ and 1360 cm⁻¹ has to be assigned to graphite-like sp^2 bonded carbon while the scattering in low frequency region around 1300 cm⁻¹ has to be interpreted in terms of scattering by sp^3 –bonded carbon. Based on this, the Raman spectra show that there is a higher concentration of sp^3 hybridized bonds in the ion beam deposited thin films than in the plasma sputtering deposited films. It is reasonable that the ion bombardment enhanced the formation of sp^3 hybridized carbon bonds. The spectra from the pure and N-doped ion beam deposited thin films are very similar, indicating that there is no significant difference in bonding state between the two samples. Generally, N-doping results in the decrease of sp^3 concentration in the film, the similar structure with N-doping might be due to the very low concentrations of nitrogen in the thin films. Further studies using synchrotron x-ray absorption spectroscopy will be performed to obtain more detailed information regarding the effect of N-doping on the bonding state of the DLC films.

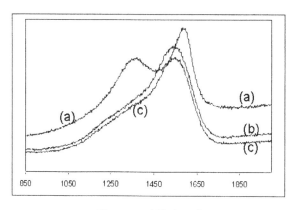

Raman shift (cm⁻¹)

Fig. 4 Raman spectra of DLC thin films (a) plasma sputtering deposited films, (b) pure and (c) N-doped ion beam deposited thin films

An average surface hardness of 18.5 GPa was obtained for the ion beam deposited DLC films on the silicon substrates with a load of 45mN. There was an increase in the hardness to 21GPa at load of 45mN on ion beam deposited nitrogen-doped DLC. Although this is in contrary to a previous published report where nitrogen doping into DLC reduces the hardness considerably[12], it is reasonable here in our case because the N-doping did not result in obvious change of sp^3 concentration in the film. sp^3 concentration is the main factor determining the hardness of DLC films, and the decrease of DLC hardness with N-doping is mainly due to the significant decrease of sp^3 concentration in the film. As the thin film structure and thickness are very similar here in our case, the increase in the hardness by N-doping might be due to the solid-solution strengthening. More experiments are being carried out to study the effect of nitrogen concentration on sp^3 concentration and hardness.

The scratch testing results based on acoustic emission and friction coefficient change show that the ion beam deposited DLC thin films have a much higher critical load (>15 N) than the plasma sputtered ones (approximately 6 N), indicating much improved adhesion of thin films to substrate due to the ion bombardment. We also observed the full scratch tracks using optical microscopy and no delamination and cracking were observed for the ion beam deposited samples when the load was up to 15 N. A typical optical micrograph of the scratched tracks is shown in Fig. 5.

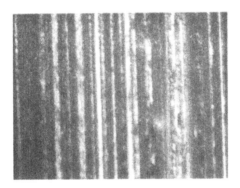

Fig. 5 Optical micrograph of the scratched tracks on the pure ion beam deposited DLC thin films

Friction and wear testing results show that there was a steady friction coefficient of 0.125 to 0.13 for the ion beam deposited pure DLC thin films and of 0.030 to 0.035 for the ion beam deposited N-doped DLC films, in consistence with the results reported by Suzuki et al.[16]. The friction keeps constant regardless of the loads. The low coefficient of friction in the N-doped DLC may be due to the formation of transfer layer during sliding wear[17]. There are intrinsic and extrinsic factors that strongly affect the friction and wear behavior of DLC thin films. The intrinsic factors include the amount of sp^2 and sp^3 hybridized bonds and concentration of hydrogen and nitrogen in the films. Extrinsic factors include surface roughness, test conditions, and the counter surface material and state. Microscopic observations of the chrome steel balls did not reveal considerable transfer layer after sliding with pure DLC thin films whereas transfer layers was observed on the balls after sliding with N-doped DLC, indicating N may enhance the formation of transfer layer to reduce the friction coefficient. The doped N may also accelerate the graphitization of DLC film during the wear test and provide a self lubrication for the sliding ball and leads to a low coefficient of friction[18]. Further experiments are being carried out to clarifying the mechanism.

2. Platelet adhesion

Platelet adhesion patterns were investigated by SEM. All samples were decorated with copper (thickness~30 nm). For each sample 25 areas of 400 μm^2 were randomly chosen on the surface contacting with PRP. Then we qualified the total platelet number N_{tot} and platelet numbers N_i in the following four morphological classes[19] (Fig. 6a)

Adhesion process is believed to run in several stages: platelet attachment to the surface, activation, pseudopodia development, spreading and aggregate formation (see Fig. 6a) [20]. The release of intracellular components from adhered or fully spread platelets (ADP, Ca^{2+}, serotonin, etc.) promotes further platelet adhesion, aggregation and finally thrombus formation[21]. One can estimate the activation of adhered platelet by their morphology. The more severe the impact of material on platelets, the more adhered cells are activated, spread or aggregated. Usual approach for quantitative investigation of platelet adhesion consists in calculation of relative index of platelet adhesion (RIPA) for different morphological classes of adhered platelets. Presented on Fig. 7 are RIPA values for carbon coatings in comparison with that of untreated PTFE. From this point of view, all carbon coatings seem more preferable for contact with platelets than untreated PTFE. Onto all coatings the numbers of slightly activated cells, spread cells and cell aggregates were lower but number of single cells was higher than that of the surface of unprocessed PTFE. It is noteworthy that the total number of platelets was also lower on carbon coated surface for all three samples. Assuming the existence of the adhesion stages, it can be suggested that platelet transition to later stages of activation was much slower on carbon coatings than on untreated PTFE.

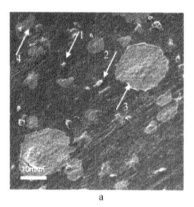 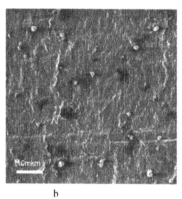

a   b

Fig. 6. Scanning electron microscope images of platelet adhesion on PTFE substrate (a) before and (b) after 1 hour of carbon deposition by the plasma sputtering:
1. Single – non-activated cells
2. Slightly activated deformed cells and pseudopodical cells.
3. Spread – fully spread platelets.
4. Aggregates –aggregates of two or more platelets.

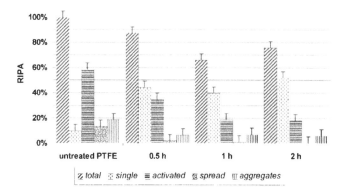

Fig. 7. Diagram representing relative platelet quantities (RIPA) for different morphological classes observed by SEM for untreated and carbon coated PTFE (time of deposition 0.5, 1, 2 hours; RIPA=100% for the total number of adhered platelets on untreated PTFE) by the plasma sputtering

CONCLUSIONS

DLC thin films were successfully deposited on to Silicon and PTFE substrates using hot wire plasma sputtering of graphite in the mixture of Ar and 5% air and by low energy ion beam deposition in a DIBD system. The property testing results show that the ion beam deposited films is very smooth and exhibits a high adhesion to substrate and a low friction coefficient when sliding with steel balls. Nanohardness measurement shows that the nitrogen-doped DLC exhibits improved surface hardness. The platelet adhesion tests demonstrate that deposition of carbon layer containing nitrogen promotes the minimization of platelet's reactions on foreign body thus considerably extending the haemocompatibility of PTFE substrate. These preliminary results show that N-doped DLC thin films are very promising for biomedical applications and further investigations are being carried out in order to understand the relationships between the processing conditions, the structure, and the mechanical, tribological and biomedical characteristics of the N-doped DLC thin films.

ACKNOWLEDGMENTS

This work is supported by the Canada Research Chair Program and NSERC. The plasma sputtering deposition in this work was conducted in the Hot Wire Reactor in the Plasma Physics Laboratory, University of Saskatchewan, which was purchased with Prof. Akira Hirose's Canada Foundation for Innovation (CFI) grant.

REFERENCES

[1] M.S. Sheu, D.M. Hudson, I.K. Loh: In "Encyclopedic Handbook of Biomaterials and Bioengineering", Part A, Vol.1, D.L. Wise, D.J. Trantolo, D.E. Altobelly, M.J. Yaszemski, J.D. Gresser, E.R. Shwartz (eds). M. Dekker (New York, 865, 1995).
[2] A. Grill: "Cold Plasmas in Materials Fabrication" (IEEE Press, New York, 1994).
[3] W.-C. Chan, M.-K. Fung, I. Bello, C.-S. Lee, S.-T. Lee: Diamond Relat. Mater. **8**, 1732 (1999).
[4] C.-F. Chen, C.-H. Shen, C.-L. Lin: Thin Solid Films, **377-378**, 326 (2000).

[5] K. Okano, T. Yamada, A. Sawabe, S. Koizumi, R. Matsuda, C. Bandis, W. Chang, B.B. Pate: Appl. Surf. Sci., **146**, 274 (1999).

[6] S.R.P. Silva, G. Amaratunga: Thin Solid Films, **253**, 146 (1994).

[7] V.N. Vasilets, A. Hirose, Q. **Yang,** A.K. Singh, R. Sammynaiken, M. Foursa, Y.M. Shulga, Characterization of doped diamond-like carbon films deposited by hot wire plasma sputtering of graphite, *Applied Physics A*, **79**, 2079 (2004).

[8] L.V. McIntire (ed): Gidelines for Blood-Material Interactions. Report of the National Heart, Lung, and Blood Institute Working Group. U.S. Department of Health and Human Services. Public Health Service. National Institutes of Health. (NIH Publication No.85-2185), 1985, Appendix 1.

[9] D. Briggs, M.P. Seah (eds.), Practical Surface Analysis by Auger and X-ray Photoelectron Spectroscopy, Wiley, Chichester (1983).

[10] J.H.Sui, W.Cai, Applied Surface Science **253**, 1050 (2006).

[11] J. Wagner, M. Ramsteiner, Ch. Wild, P. Koidl: Phys. Rev. **40**, 1817 (1989).

[12] M.Guerino, M.Massi, H.S Maciel, C.Otani, R.D.Mansano, P.verdonck, J.Libardi, Diamond and Related Materials, 13, 316 (2004).

[13] A. Laskarakis, S. Logothetidis, M Gioti: Phys. Rev. B, **64**, 125419 (2001).

[14] J.H. Kaufman, S. Metin, D.D. Saperstein: Phys. Rev. B, **39**, 13053 (1989).

[15] Y. Liu, C Jiaa, H. Do: Surf. Coat. Technol., **115**, 95 (1999).

[16] M.Suzuki, T.Watanabe, A. Tanaka and Y. Koga, Diamond and Related Materials, **12**, 2061 (2003).

[17] M. Jarratt, J. Stallard, N. M. Renevier and D. G. Teer, Diamond and Related Materials, **12**, 1003 (2003).

[18] Yin-Yu Chang, Da-Yung Wang, Chi-How Chang and WeiTe Wu, Surface and Coating Technology, **184**, 349 (2004).

[19] S.L. Goodman: J.Biomed. Mater. Res. **45**, 240 (1999).

[20] S.L. Goodman: K.S. Tweden, M. Albrecht: J.Biomed. Mater. Res. **32**, 249 (1996).

[21] L.M. Waples, O.E. Olorundare, S.L. Goodman, Q.J. Lai, R.M. Albrecht: J.Biomed. Mater. Res. **32**, 65, (1996).

NANOSTRUCTURED NITRIDE SURFACE VIA ADVANCED PLASMA NITRIDING AND ITS APPLICATIONS

Sehoon Yoo, Yong-Ki Cho, Sang Gweon Kim, Sung-Wan Kim
Korea Institute of Industrial Technology
Incheon, 406-840, Korea

ABSTRACT

Nanostructured nitride layers on steel surface were developed via our advanced plasma nitriding, ATONA(atomic nitriding & application) process. The grain size of nitrided layer was ranged from 30 to 80 nm. Unlike conventional ion nitriding(CIN), ATONA adopted mesh-type, double screens around components and plasma was generated on the screen rather than on the component surface. Another feature of ATONA was relatively low working pressure(below 0.1 torr) and temperature(450°C) as compared with CIN. ATONA-treated steel specimens possessed lower compound layer thickness than CIN-treated specimens, which lowered the possibility of surface failures. The generated nanostructured nitride surface enhanced the adhesion of diamond like carbon (DLC) with steel surface up to 70 N, which was about at least 3 times as high as that of conventional DLC coatings. CrN coating deposited on a polished ATONA-treated specimen also had high adhesion of 80 N. Carbon nanotubes(CNTs) were also successfully deposited on the nano-sized nitride layers by ATONA without additional catalyst deposition.

INTRODUCTION

Most failures for mechanical components occur on surfaces such as wear, corrosion, fatigue fracture, fretting fatigue, etc. Therefore, optimization of the surface microstructure and chemistry is an effective approach to eliminate the failures of mechanical parts. One of the effective modifications of surface microstructure is surface nanocrystallization. Surface nanocrystallization increased surface hardness[1], physical properties[2], and wear and corrosion properties[3, 4]. Surface chemistry modification techniques to enhance surface reliability are nitriding and ceramic coatings[5]. Among the techniques, plasma nitriding technique has become industrially important because of lower part distortion and less emission of polluting gases than gas nitriding[5]. Plasma (or ion) nitriding is a surface hardening technique that utilizes glow discharge to introduce elemental nitrogen into a workpiece. Recently, plasma assisted nitriding(PAN) that utilizes atomic nitrogen has been widely studied because PAN is able to produce compound-layer-free nitriding layers on steels and thick nitriding layers on non-ferrous metals[6-8]. However, PAN has not been actively applied to industries because treatment area is so small that it requires longer time for mass production than conventional ion nitriding (CIN).

In this study, a surprisingly highly-scalable technique has been developed for nanostructured surface nitriding layers. The nano-sized nitriding surface was formed via our novel plasma nitriding techniques, ATONA(ATOmic Nitriding & Applications) process. The ATONA process adopted mesh-type, double screens around workpieces and negative voltage was applied on the double screens under nitrogen/hydrogen gas environment. Unlike conventional ion nitriding(CIN) of steels, ATONA utilized relatively low working pressure (below 0.1 torr) and low temperature (450°C). Although ATONA utilized low pressure, high plasma density was able to be achieved from restricted electrons between negatively charged double screens. Since ATONA modified the nitriding chamber for CIN process, the treatment area of ATONA process was as large as CIN.

The combination of microstructure and chemistry modifications by ATONA process provided unique properties of the nitrided surface. In this article, the microstructures and surface properties of

ATONA- and CIN-treated steel samples were observed with optical microscope, secondary electron microscope (SEM), and surface hardness testing. Besides its use of surface hardening, the ATONA method was used for producing underlayers for subsequent deposition of diamond like carbon(DLC) and ceramic coatings. Such coatings have been applied to the mechanical components of automobiles, machineries, tools and dies to provide enhanced wear- and corrosion-resistance. To ensure surface wear and corrosion properties, adhesion between the coatings and workpieces should be high enough to resist severe wear environment. Therefore, the adhesions of DLC and CrN coatings on the nanostructured nitride layer by the ATONA technique were analyzed with scratch testing. In addition, the nano-sized grains by ATONA were also used as seeds for carbon nanotube(CNT) growth that was observed with scanning electron microscopy(SEM).

EXPERIMENTAL

Specimens for ATONA and CIN process were AISI 1045 steel and the chemical composition is shown in Table I. The specimens experienced quenching from 870°C and tempering at 160°C. The sample was then polished with SiC paper followed by fine polishing with 1 μm diamond pastes. The polished AISI 1045 steel specimens were ultrasonically cleaned in acetone.

After cleaning, the AISI 1045 specimens were placed on a sample stage in plasma nitriding system as shown in Figure 1. With the plasma nitriding system, both ATONA and CIN were carried out by applying DC power on the screen or on the stage, respectively. Base pressures for both ATONA and CIN process were 6×10^{-3} torr. Temperatures for ATONA and CIN were 450°C and 550°C, respectively. Before main process, specimens were sputter-cleaned with Ar/H$_2$ plasma at the pressure of 1.25 torr for 1 hr. During the sputtering cleaning, bias voltage of -500 V was applied on the stage. After the sputter-cleaning, N$_2$ and H$_2$ gases were backfilled to the nitriding chamber and 10-20 kW of power was applied on the screen for ATONA process and on the stage for CIN process. The working pressure was 0.07 torr for ATONA and 2 torr for CIN. The ratio of N$_2$ to H$_2$ was 1:3.

DLC was coated onto ATONA treated AISI 1045 specimens with plasma enhanced chemical vapor deposition(PECVD). Before the DLC coatings, specimen surface was sputter-cleaned with Ar plasma. Processing gases for the DLC deposition were Ar, H$_2$, NH$_3$, TMS, and CH$_4$. Deposition pressure of DLC was 2 torr. Pulsed DC power was applied on the specimens during the DLC deposition. CrN was also deposited onto ATONA specimens with arc ion plating(AIP) technique. The working temperature and pressure were 300°C and 7.5 torr, respectively. Arc power was applied on Cr target under N$_2$ gas for the CrN coating.

CNTs were deposited on both ATONA- and CIN-treated nitride layers via thermal chemical vapor deposition. Specimens were loaded in a fused silica reaction tube within a horizontal tube furnace. Then, the tube was purged with H$_2$/N$_2$ gas mixture for 30 min and was heated at 12.5°C/min to 600°C. At 600°C for 40 min, the H$_2$/N$_2$ gas mixture reduced native iron oxide formed on the nano-size nitride layer. After the 40 min of the reduction, C$_2$H$_2$ gas was introduced to the H$_2$/N$_2$ gas mixture for CNT growth for 5 min. After the CNT growth, tube was cool down to RT.

Microstructure was observed by secondary electron microscope (SEM, HITACH S-4300). Surface hardness and case depth were determined with micro Vickers hardness tester(Future Tech FM-7). Indentation cracking test was carried out with Rockwell hardness tester(Matsuzawa DXT-3) with load of HRC150. Adhesion was analyzed with a scratch tester (CSEM) equipped with a diamond tip. The normal load was increased from 0 to 100 N as a loading rate of 100 N/min. The scratching speed was 10 mm/min. During the scratch test, acoustic emissions and friction forces were continuously monitored. The microstructure of the scratch track was observed with an optical microscope. Friction coefficient was measured by pin on disc method. For pin on disc test, 6mm-diameter steel ball was used. The test temperature was room temperature and humidity was about 50%. The friction coefficient was measured up to 500 m. Rotating speed was 3 m/s and test load was 10 N.

Table 1. Chemical composition of AISI 1045 carbon steel

Element	C	Mn	Si	P	S
Wt%	0.42-0.48	0.6-0.9	0.15-0.35	≤0.03	≤0.03

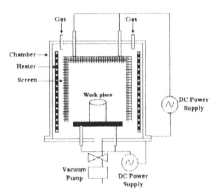

Figure 1. Schematics of nitriding system for ATONA and CIN process

RESULT AND DISCUSSION

As shown in the secondary electron image of Figure 2(a), 2 hrs of ATONA process resulted in the formation of nano-sized grains on the AISI 1045 steel. The grain size was ranged from 30 to 80 nm. On the other hand, 2 hrs of conventional ion nitriding (CIN) yielded 1-2 μm grains on the surface of AISI 1045 steel as shown in Figure 2(b). Figure 3 shows cross-sectional micrographs of the AISI 1045 steel after 2 hrs of ATONA and CIN treatments. Generally, most nitriding surface of steel has compound layer that was composed of γ' (Fe_4N) and/or ϵ ($Fe_{2-3}N$) intermetallics[5]. On the compound layer, most failures occur when the compound layer is thick or porous. Therefore, light grinding (to remove the compound layer) following the nitriding process is often required to eliminate the failure on the compound layer. As shown in Figure 3, thickness of the compound layer was about 500 nm for the ATONA processed AISI 1045 steel. On the other hand, the thickness of the compound layer of CIN specimen was over 20 μm. Since compound layer of ATONA specimen was very thin, failure may have not been as severe as CIN specimen.

To investigate the failure on the compound layer, indentation cracking test was carried out as shown in Figure 4. After indentation cracking test of the CIN specimen, extended cracks were found near the circular indentation. On the other hand, no crack was found near indentation for ATONA specimen. These results indicated that mechanical failure did not occur on ATONA specimen because it had very thin compound layer.

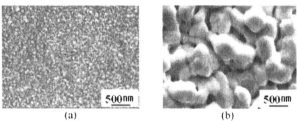

(a)   (b)

Figure 2. Scanning micrographs of AISI 1045 steel after 2 hr treatment of
(a) ATONA and (b) CIN.

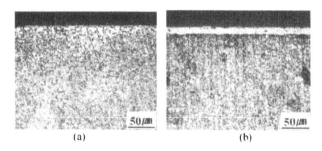

(a)   (b)

Figure 3. Cross-sectional SEM images of AISI 1045 steel after 2 hr treatment of
(a) ATONA and (b) CIN.

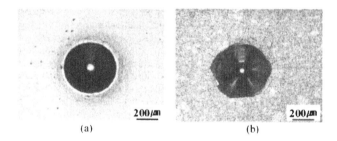

(a)   (b)

Figure 4. Indentation cracking test results of (a) ATONA treated AISI 1045 specimen and (b) CIN
treated AISI 1045 specimen.

ATONA process also enhanced nitrogen diffusion rate to the matrix. Immediately below the compound layer, diffusion layer(or diffusion zone) is formed by nitrogen diffusion. The diffused nitrogens are interstitially or substitutionally located in the iron matrix or form nitrides with nitride-forming elements at grain boundaries or even within grains. Such nitrogen diffused zone right under the compound layer also increases surface hardness. This diffusion zone, however, can not be clearly observed with microscope. Therefore, the nitrogen diffusion layer is determined indirectly by measuring hardness with depth. Hardness depth profiles of ATONA- and CIN-treated steel sample were measured with Vickers hardness tester and shown in Figure 5. Surface hardness of ATONA-treated AISI 1045 steel was about 810 Hv which was higher than that of CIN-treated AISI 1045 steel. In addition, overall hardness of ATONA-treated specimen had higher hardness than CIN-treated specimen. This high hardness of ATONA specimen resulted from fast nitrogen diffusion to the matrix. Such fast nitrogen diffusion of ATONA process compared with CIN may have been due to the surface nano-grains and thin compound layer. Adsorption of nitrogen may have increased due to the large portion of grain boundaries of the nanostructured nitride layer since chemical affinity was increased by the long grain boundaries. Therefore, the amount of adsorbed nitrogen may have been much higher on the ATONA sample than on the CIN sample, which caused high concentration gradient of nitrogen. In addition, thin compound layer of ATONA treated sample may have caused fast nitrogen diffusion because diffusion of nitrogen was slower in the compound layer than in the matrix[9]. In our case, the compound layer thickness was about 500 nm after 2hrs of ATONA treatment, which was at least 40 times lower than that of 2hr-CIN-treated sample. Therefore, nitrogen diffusion rate on the ATONA treated AISI 1045 steel was much higher than that on CIN treated AISI 1045 steel.

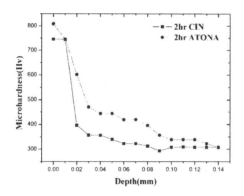

Figure 5. Hardness profiles of AISI 1045 steels after 2hrs of ATONA and CIN process.

The nanostructured surface layers by ATONA were also used as underlayers for DLC coatings to increase adhesion. The adhesion of DLC coatings is generally poor because of high residual compressive stress of the DLC films and stability of the carbon-carbon bonds, which reduce the chemical affinity to different substrate materials. Although many attempts to reduce the residual stress has been successful, reducing the residual stress of DLC films caused other mechanical problems[10]. Therefore, most adhesion related researches on DLC has been focused on the increase of chemical affinity by modifying the substrate surface, i.e. Ar sputtering of the substrate prior to deposition or introducing an intermediate layer which is chemically compatible to both carbon and substrate materials[11]. Among the intermediate layers for adhesion improvement, Si, Ti and their carbide layers have been widely studied for last decade[12-16]. Other metal such as Ni, Cr, Zr, Al or W were also investigated as a possible intermediate layer[11, 17].

However, industries still require higher adhesion of DLC for the application of severe mechanical applications. In our case, the nanostructured nitride layer was used to increase adhesion of DLC coatings. To understand the effect of nanostructured surface layer on the adhesion, DLC with thickness of 1.2 μm was deposited onto the nanostructured nitride layers by ATONA process. After DLC deposition, the adhesion between DLC and nitrided surface was measured by scratch testing. Figure 6 shows optical micrograph of the scratch track after scratch testing. Critical load at which failure occurred was 70N, which was at least 3 times larger than conventional DLC coatings. The high adhesion was appeared to be due to large portion of grain boundaries of the nanostructured nitride layer, which increased surface energy. To lower the surface energy, the nanostructured surfaces tended to chemically combine well with depositing species. Friction coefficient was also measured for the DLC coatings deposited on the nanostructured nitriding layer and was shown in Figure 7. The friction coefficient of DLC measured was 0.05 and the value was very stable up to 500 m.

Figure 6. Optimcal micrographs of scratching track of DLC films deposited onto nanostructured nitride layer by ATONA process.

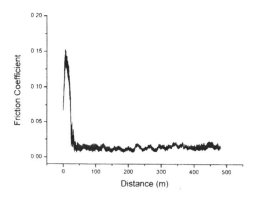

Figure 7. Friction coefficient of DLC films deposited onto nanostructured nitride layer by ATONA process.

The nanostructured nitride grains by ATONA process were used as an underlayer for subsequent CrN coatings. The combination of nitriding and hard coatings has been recently introduced to fulfill the increasing demand of better performance in severe operation environment in the field of automobiles, machineries, tools, and molds[18, 19]. Such dual coating is also required high adhesion for mechanical applications. To understand the adhesion properties, CrN was deposited onto bare steel, ATONA treated steel and polished ATONA treated steel. After deposition, adhesion was measured with scratch test and was shown in Figure 8. Unlike DLC adhesion results, critical load of CrN on ATONA specimen was 10 N lower than that on bare steel (Figure 8(b)). However, when ATONA specimen was polished with alumina slurry for 10 sec (Figure 8(c)), the adhesion was surprisingly increased to 80 N, which was twice higher than that of bare steel. Recent research reported that surface modification by grinding with diamond wheel increased adhesion of DLCs and it discussed the high adhesion was due to the high chemical affinity of modified layer by grinding[20]. The high CrN adhesion on the polished ATONA specimen had not been clearly understood at the moment but it may have been related to the modified layer of the polished ATONA specimen.

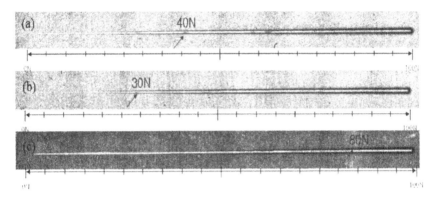

Figure 8. Optimcal micrographs of scratching track of CrN coatings deposited onto (a) polished steel surface, (b) ATONA specimen, and (c) polished ATONA specimen with alumina slurry for 10sec.

The nanostructured nitride layer was also used as CNT underlayer. Generally, CNT growth requires transition metal deposition. In our case, instead of transition metal deposition, the nanostructured nitride was used as underlayers for the CNT growth. After heating in $C_2H_2/H_2/N_2$ gas mixture at 600°C for 10min. CNTs were successfully formed on the ATONA treated nanostructured nitride surface as shown in Figure 9(a). On the other hand, thick fibers were observed on the CIN treated sample but no CNTs were deposited (Figure 9(b)). For the ATONA sample, high activity of nanosized grains may have enhanced absorption and decomposition of the carbon-bearing gas. In addition, grain boundaries of iron nitride may be favorable sites for CNT growth. Huang et al.[21] reported CNTs were formed only at grain boundaries when CNT were grown on Fe-N films when CNT was formed by PECVD. Such report indicated that long grain boundaries of nanostructured ATONA specimen increased the possibility of CNT formation.

Figure 10 shows some of applications that ATONA has been successfully used in. Most often they have applied to automobile parts. precision tools, and dies resulting in improved wear life and mechanical reliability.

(a)   (b)

Figure 9. Scanning electron micrographs of CNTs deposited onto (a) ATONA treated sample and (b) CIN treated sample.

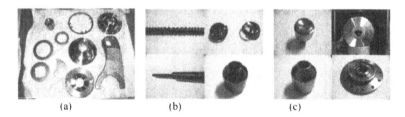

(a)   (b)   (c)

Figure 10. Mechanical parts treated with (a) ATONA process, (b) DLC on ATONA treated surface, and (c) CrN coatings on ATONA treated surface.

CONCLUSION

Nanostructured nitride layer of AISI 1045 steel with diameters of 30-80 nm was produced with advanced plasma nitriding, ATONA process. During ATONA nitriding, plasma was generated on double screens around components under low working pressure (below 0.1 torr) and low temperature (450°C). The compound layer after 2hrs of ATONA process was about 500 nm which enhanced mechanical reliability. The nitriding rate during ATONA was almost 3 times faster than that of conventional plasma nitriding due to fast diffusion through the nano-sized grains. The generated nano-sized nitriding surface improved adhesion of diamond like carbon (DLC) with steel surface up to 70 N, which was about at least 3 times as high as that of conventional DLC coatings. CrN coatings were also deposited on the ATONA treated nitride layer and showed enhanced adhesion of 80 N. CNTs were successfully generated on the nano-sized nitride surface without additional catalyst deposition.

ACKNOWLEDGEMENT

Financial support for this work was provided by the Korea Ministry of Planning and Budget foundation through the Korea Institute of Industrial Technology (KITECH).

REFERENCES
[1]K. Lu and J. Lu, Nanostructured surface layer on metallic materials induced by surface mechanical attrition treatment. *Mater. Sci. Eng. A*, **375-377**, 38-45 (2004).
[2]W. P. Tong, N. R. Tao, Z. B. Wang, J. Lu, and K. Lu, Nitriding iron at lower temperatures. *Science*, **299**, 686-88 (2003).

[3]X. Y. Wang and D. Y. Li, Mechanical, electrochemical and tribological properties of nano-crystalline surface of 304 stainless steel, *Wear*, **255**, 836-45 (2003).

[4]T. Wang, J. Yu, and B. Dong, Surface nanocrystallization induced by shot peening and its effect on corrosion resistance of 1Cr18Ni9Ti stainless steel, *Surf. Coat. Technol.*, **200**, 4777-81 (2006).

[5]J. R. Davis and D. Associates, Surface Hardening of Steels, ASM International, Materials Park, OH (2002).

[6]T. Czerwiec, H. Michel, and E. Bergmann, Low-pressure, high-density plasma nitriding: mechanisms, technology and results, *Surf. Coat. Technol.*, **108-109**, 182-90 (1998).

[7]A. Leyland, K. S. Fancey, A. S. James, and A. Matthews, Enhanced plasma nitriding at low pressures: A comparative study of d.c. and r.f. techniques, *Surf. Coat. Technol.*, **41**, 295-304 (1990).

[8]N. Renevier, H. Michel, T. Czerwiec, and P. Collignon, Diagnostic of arc discharges for plasma nitriding by optical emission spectroscopy, *Surf. Coat. Technol.*, **98**, 1400-05 (1998).

[9]H. Seong, B. Yoo, and K. Cho, The Formation Rate and Activation Energy of Diffusion Layer and Compound Layer in Ion-Nitriding, *Trans. Korean Soc. Mech. Eng.*, **8**, 476-80 (1984).

[10]K.-R. Lee, Y.-J. Baik, K. Y. Eun, and S. Han, Precursor gas effect on the structure and properties of diamond-like carbon films, *Diamond Rel. Mater.*, **3**, 1230-34 (1994).

[11]K.-R. Lee, K. Y. Eun, I. Kim, and J. Kim, Design of W buffer layer for adhesion improvement of DLC films on tool steels, *Thin Solid Films*, **377-378**, 261-68 (2000).

[12]C. Dumkum, D. M. Grant, and I. R. McColl, A multilayer approach to high adhesion diamond-like carbon coatings on titanium, *Diamond Rel. Mater.*, **6**, 802-06 (1997).

[13]R. A. Erck, F. A. Nichols, and J. F. Dierks, Pull-test adhesion measurements of diamondlike carbon films on silicon carbide, silicon nitride, aluminum oxide, and zirconium oxide, *J. Vac. Sci. Technol. A*, **12**, 1583-86 (1994).

[14]A. Erdemir, M. Switala, R. Wei, and P. Wilbur, A tribological investigation of the graphite-to-diamond-like behavior of amorphous carbon films ion beam deposited on ceramic substrates, *Surf. Coat. Technol.*, **50**, 17-23 (1991).

[15]A. Grill, B. Meyerson, and V. Patel, Interface modifications for improving the adhesion of a-C:H films to metals, *J. Mater. Res.*, **3**, 214-17 (1988).

[16]H. Ronkainen, J. Vihersalo, S. Varjus, R. Zilliacus, U. Ehrnstén, and N. P., Improvement of a-C:H film adhesion by intermediate layers and sputter cleaning procedures on stainless steel, alumina and cemented carbide, *Surf. Coat. Technol.*, **90**, 190-96 (1997).

[17]H. Freller, A. Hempel, J. Lilge, and H. P. Lorenz, Influence of intermediate layers and base materials on adhesion of amorphous carbon and metal-carbon coatings, *Diamond Rel. Mater.*, **1**, 563-69 (1992).

[18]S.-C. Lee, W.-Y. Ho, and W.-L. Pao, Process and properties of CrN coating deposited on plasma nitrided high-speed steel, *Surf. Coat. Technol.*, **73**, 34-38 (1995).

[19]K. Kondo, K. Ikenaga, Y. Funaki, and H. Takeuchi, Adhesion Strength between D.C. Plasma Nitrided Steels and CrN Films, *Heat Treat. Mater.*, **2006**, 265-68 (2006).

[20]H. Ohmori, K. Katahira, J. Komotori, M. Mizutani, F. Maehama, and M. Iwaki, Investigation of Substrate Finishing Conditions to Improve Adhesive Strength of DLC Films, *CIRP Annals Manu. Technol.*, **54**, 511-14 (2005).

[21]L. Huang, S. P. Lau, D. G. McCulloch, W. H. Zhong, C. Q. Sun, Y. Q. Fu, and B. K. Tay, Self-organized carbon nanotubes grown at the grain boundary of iron-nitride, *Carbon*, **43**, 651-73 (2005).

Author Index

Lightning Source UK Ltd.
Milton Keynes UK
UKOW06n1403150416

272324UK00001B/35/P